T0269080

In recent years there has been an explosion of interest in the study of molecular crystals and their applications in optics and electronics. This advanced textbook describes their chemical and physical structure, their optical and electronic properties, and the reactions between neighbouring molecules in crystals. The author has taken into account research areas which have undergone extremely rapid development since the first edition was published in 1987. For instance, it features the applications of molecular materials in high-technology devices. There is also an entirely new chapter on C_{60} and organic non-linear optic materials. The level of treatment is aimed at first-year postgraduates or workers in industrial research laboratories wishing to gain insights into organic solid-state materials. It is also suitable for special topics in final-year undergraduate courses in chemistry, physics and electronic engineering.

Molecular crystals

J. D. WRIGHT *University of Kent at Canterbury*

Molecular crystals

SECOND EDITION

CAMBRIDGE UNIVERSITY PRESS
Cambridge, New York, Melbourne, Madrid, Cape Town, Singapore,
São Paulo, Delhi, Dubai, Tokyo, Mexico City

Cambridge University Press
The Edinburgh Building, Cambridge CB2 8RU, UK

Published in the United States of America by Cambridge University Press, New York

www.cambridge.org
Information on this title: www.cambridge.org/9780521477307

First published 1987
First paperback edition 1989
Second edition 1995

A catalogue record for this publication is available from the British Library

Library of Congress Cataloguing in Publication Data

Wright, J.D. (John Dalton), 1941-
  Molecular crystals / J.D. Wright. - 2nd ed.
    p. cm.
  ISBN 0-521-46510-9. - ISBN 0-521-47730-1 (pbk.)
  1. Molecular crystals. I. Title.
QD921.W72 1995
548 - dc20 94-16148 CIP

ISBN 978-0-521-46510-6 Hardback
ISBN 978-0-521-47730-7 Paperback

Contents

Preface to second edition

Since this text was first published there have been significant advances in the applications of molecular crystalline materials, and the whole area of C_{60} chemistry has developed on an enormous scale. The major features of this second edition are a significantly increased coverage of these areas. In the case of non-linear optic materials and C_{60}, special-topic chapters have been added at the end of the text. Both these areas draw heavily on material already described in several of the existing chapters. They therefore form excellent examples of the value of this background material. Other molecular electronics applications covered include high-technology applications of organic dyes (e.g. WORM optical data storage devices, colour microfilters for liquid crystal displays), the use of organic materials in electrophotography (photocopiers and laser printers), electroluminescent displays and photovoltaic solar cells. Significantly, much of the necessary background in ideas and materials was already covered, and these sections cross-reference many of the chapters of the first edition.

There have also been useful advances in techniques, and scanning probe microscopies are now included in chapter 4, while positron annihilation is now described more concisely, reflecting its diminishing use. Mention is also made of the growing use of crystallographic databases to establish structural trends and to identify structure-determining interactions. The application of this improved understanding of the interactions determining molecular assembly to the design of self-assembling molecules is also discussed in chapter 3, and examples of the use of these ideas are included in the discussions of gas-sensor and non-linear optical materials.

Apart from correction of minor errors, I have resisted the temptation to expand the bulk of the material from the first edition in the interests of retaining a concise format which provides a broad introduction to the subject. Although it has been argued that it is simply not possible to convey such a wide field adequately in such a concise manner, evidence from both academic staff and students from many institutions has suggested that this approach is useful to gain a foothold in the subject which is not provided by larger, more comprehensive volumes. Similarly, line drawings of crystal structures etc. continue to be used rather than the excellent computer-generated molecular graphics which are now widely available, since students have some chance of sketching an approximation to the line drawings themselves to illustrate a point. The more advanced the graphics, the less likely it is that students will attempt this.

I thank the many colleagues and students who have provided helpful feedback contributing to this edition, and look forward to the continuation of this valuable source of new insights.

John D. Wright
January 1994

Preface to first edition

Many of the physical and chemical properties of molecular crystals are very different from those of other solid-state materials. There are also extensive possibilities for varying these properties by altering the electronic and geometric structure of the molecules using the synthetic skills developed by organic chemists. The exploitation of these possibilites in new material applications is entering a rapid-growth phase, supported by a very strong research base world-wide. However, in the teaching of solid-state chemistry and physics, molecular crystals are seriously neglected. Most of the current solid-state chemistry and physics textbooks at the level of final-year undergraduate/first-year postgraduate studies scarcely mention molecular crystals. There are several excellent advanced texts, but these cover limited areas of the subject and by their detailed nature tend to be difficult for newcomers, as well as expensive. This situation hinders the effective training of the growing number of scientists who wish to work with molecular crystalline materials. My objective in writing this book has been to fill this gap.

The range of material covered in the book is very wide, as is the range of inherent difficulty of the underlying concepts, and there are several places where details of experimental methods or of mathematical developments of theories (e.g. of exciton theory) would have made the text inordinately long or complex. In these cases the supporting references provided with each chapter should be particularly useful. These references are representative but by no means exhaustive. To have referenced all the elegant examples from the research literature would be impossible in a text of this level, and I apologise to those whose work is not cited. Omissions and variations in level of treatment are inevitable but not deliberate.

The book is based on over ten years of teaching an undergraduate course

on molecular crystals to final-year students, and I would like to acknow-
ledge the challenging and perceptive questioning of these students, which
has often forced me to examine and clarify my own thinking. I would also
like to express my thanks to scientific colleagues from many countries for
their contribution to my understanding of this field, to my research students
and my family for their forbearance during my preoccupation with writing
the book, and to John Couves for his helpful and critical comments, as a
first-year postgraduate, on the manuscript.

November 1986

The publication of the paperback edition (1988) realised my original
objective of a text accessible to students for their own libraries.

The citation of original references wherever possible has led to the
omission of any reference to the following advanced texts: M. Pope and
C.E. Swenberg, *Electronic Processes in Organic Crystals*, Oxford: Claren-
don Press, 1982; H. Meier, *Organic Semiconductors*, Weinheim: Verlag
Chemie, 1974; E.A. Silinsh, *Organic Molecular Crystals*: *Their Electronic
States*, Berlin: Springer, 1980; K.C. Kao and W. Hwang, *Electrical
Transport in Solids*, Oxford: Pergamon, 1981. These texts have had a
significant influence on my own perceptions of the subject and can be
strongly recommended for those wishing to extend their reading, in
addition to those texts covering related areas which are cited in their own
contexts.

1988
John D. Wright

1

Purification and crystal growth

Almost all the physical properties of molecular crystals can be influenced by crystal purity and quality. Thus, if experimental measurements of such properties as energy and electron transfer, molecular motion or solid-state chemical reactions in molecular crystals are to be interpreted in terms of theoretical models, it is vitally important that these measurements are made as far as possible on single crystals which are as pure and perfect as possible. Unfortunately, the relatively weak short-range intermolecular forces in molecular crystals frequently render this objective difficult to achieve. For example, impurity molecules may often be incorporated into the lattice with the rather small, unfavourable enthalpy caused by localised lattice strain around the impurity site offset by a favourable entropy of mixing. Also twinning, dislocations, disorder and even strain-induced inclusions of regions of different lattice structure can occur relatively easily, with minor effects on the lattice energy but possibly major effects on physical properties.

1.1 Purification

Conventional chemical standards of purity are of little value in the field of organic solids. For example, in a three-dimensional lattice for a material of 99.9% purity, in any given direction one molecule in ten is likely to be an impurity molecule – clearly undesirable compared with any model assuming an extended regular lattice. Starting materials for molecular crystals, whether commercial or synthesised in the laboratory, must therefore be further purified in almost every case prior to crystal growth.

Purification methods fall into two classes: physical methods based on phase separation, including distillation, sublimation, recrystallisation, chromatography and zone-refining; and chemical methods, in which impurities that cannot be removed by physical methods are first selectively converted to materials which can be more easily separated. Since many organic molecules are photosensitive or readily oxidised, it is frequently advantageous to carry out purification processes in the dark and under inert atmospheres.

Distillation and recrystallisation

Distillation and recrystallisation are useful initial steps in purification. The risk of thermal decomposition is reduced by vacuum distillation, and repeated fractional distillation can lead to high purity. Recrystallisation has the advantage of lower temperature, with less likelihood of thermal decomposition, but requires purified solvents and may lead to incorporation of solvent in the solid product. Also, the concentrations of impurities which are less soluble than the material being purified may actually be increased rather than reduced by recrystallisation (e.g. anthracene as an impurity in phenanthrene). In such cases, chromatographic methods may be more effective or chemical methods may be required.

1.1. Continuous-chromatography apparatus.

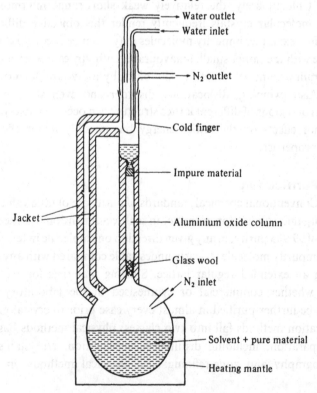

Continuous chromatography

Figure 1.1 shows a typical system for continuous chromatography[1], in which only a small volume of solvent is required for preliminary purification of a large amount of material. The adsorbent (frequently alumina or silica) and solvent are chosen according to the chemical nature of the material and the likely impurities present. The degree of purification achieved in such a system is limited owing to the large amount of material passed through a relatively small column, and the fact that impurities which are only weakly held by the adsorbent will not be effectively removed. Also, the purified material collects in the boiling solvent at a relatively high temperature, requiring careful exclusion of light and oxygen if undesirable reactions are to be avoided. Higher purity is obtained by non-continuous chromatography. The progress of the material and impurities through the column may be followed in many cases by occasional examination of fluorescence bands in the column under ultra-violet illumination. More elaborate and expensive techniques such as high-pressure liquid chromatography combine the advantages of high purity and high speed, although these are best suited to relatively small quantities of material.

Sublimation

Processes such as chromatography or recrystallisation, which may lead to incorporation of solvent in the molecular crystal, are best followed by further purification by sublimation. This eliminates all traces of volatile solvents and also involatile impurities such as any thermally polymerised material. In the simplest sublimation apparatus, the material is heated in an evacuated chamber containing a cold finger on which the vaporised material is condensed. Impurities with vapour pressures similar to the material being purified will not be removed using this apparatus, and in such cases gradient sublimation is preferred, in which the material is sublimed down a gradual temperature gradient along a horizontal tube. The temperature gradient is produced by a furnace containing two or more zones of different temperature, and the tube may be continuously evacuated or sealed following initial evacuation. Entrainer sublimation is carried out in a slow flow of inert gas (e.g. nitrogen) usually at atmospheric pressure (figure 1.2). Volatile impurities are usually best separated by entrainer sublimation or in continuously evacuated systems, although in these conditions relatively involatile constituents may be swept down the tube by other vapour species, leading to contamination or poor separation. This effect may be minimised by subliming the material through a layer of adsorbent such as charcoal or glass wool to condense the less volatile materials. After gradient sublimation has been carried out, the tube is cut into zones containing material of different purity, and the material from the purest zones may be used as starting material for further sublimation stages. The process may need to be carried out several times to achieve reproducible material of a high degree of purity. The ultimate degree of purity will depend on the temperature gradient and rate of sublimation (ideally both low) as well as on the thermal stability of the material and the extent to which trace impurities of similar volatility were present in the original sample.

Zone-refining

A more convenient method for carrying out many successive cycles of purification for materials which can be melted without decomposition is zone-refining. This process, developed in 1952 by W. G. Pfann[2] for purification of inorganic semiconductor materials, involves repeated passage of a narrow molten zone along a rod of the material to be purified. Impurities which are more soluble in the melt than in the host solid are carried along the rod in the direction of movement of the molten zone. Impurities which are less soluble in the melt than in the host solid will tend to solidify first as the zone progresses, and thus move in a direction opposite to that of the molten-zone movement, in steps whose maximum length is equal to that of the molten zone. More passes are therefore needed to move the latter impurities a given distance along the rod than for the former. If the relative solubility of an impurity in the solid and molten phases is known, it is possible to develop theoretical models for calculation of the progress of

1.2. Entrainer-sublimation furnace.

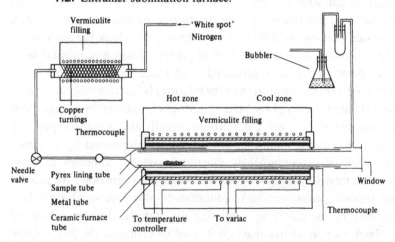

the purification and the resulting distribution of impurity along the length of the rod at various stages of zone-refining.

In practice it is not possible to use such calculations to predict the number of zone passes needed to achieve a certain degree of purity in a molecular solid, since there are usually several impurities present whose identity, concentration and solubility are unknown. Instead, automatic zone-refining apparatus is used to pass a large number of molten zones through the sample over several days. For a sample containing several impurities, some more and some less soluble in the melt than in the host solid, a typical impurity distribution along the length of the zone-defined rod is shown in figure 1.3. Segregation of impurities at the ends of the rod may be observed as a discoloration of the material in many cases, while examination under an ultra-violet light source, which excites fluorescence is a more sensitive guide to purity (see chapter 4). If material from the central portion of the rod is not of adequate purity, the central regions of several rods may be combined and subjected to further zone-refining until the limiting purity level is attained for the new sample. The ultimate limiting purity is determined by the point at which the impurity concentration in the end zones is high enough for the solid crystallising at the zone edge to be of the same composition in consecutive zone passes. This point is determined by the solubility ratio of the impurity in the solid and melt, and by the ratio of the total length of the sample to the length of the molten zone. Indirectly, the rate of zone movement is also important. If the zone moves too rapidly, impurity molecules are trapped at the advancing solidifying surface, and in extreme cases molten regions may be entrapped also. With slow zone movement through materials of high purity, back-diffusion of impurity molecules may become the chief factor limiting purity. Photochemical reactions of the material being zone-refined may be avoided by carrying out the operation in the dark, but any thermal reactions will inevitably limit ultimate purity.

Figure 1.4 shows a typical automatic zone-refining apparatus capable of passing several molten zones simultaneously through the sample. Since

1.3. Impurity distribution along a zone-refiner tube.

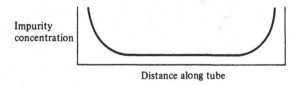

Impurity
concentration

Distance along tube

most molecular solids are poor thermal conductors, the sample tube is usually 1 cm or less internal diameter. Wider tubes require more powerful heaters to maintain the whole cross-sectional area of the tube in a molten state, resulting in wider zones at the outer tube surface. Very narrow tubes are also inadvisable since any bubbles forming in the molten zone may lead to discontinuities in narrow samples. The starting material is prepared by loading a long tube, sealed at one end, with powdered material and then evacuating the tube. An inert gas (e.g. argon) is admitted to a pressure of about 10 cmHg before the sample is melted, from the bottom upwards, and sealed. Melting and sealing should be carried out rapidly to avoid excessive sublimation of the material, and it is advisable to use a loose plug of glass wool in the upper end of the tube to prevent accidental contamination of the vacuum line during this operation. To commence zone-refining, the sample should be re-melted, from the top downwards, and loaded into the pre-heated zone-refining apparatus while still completely molten to avoid risk of the tube cracking.

1.2   Crystal growth

Although some of these purification techniques (e.g. sublimation) are capable of producing good-quality single crystals, crystal growth is

1.4. A zone-refiner.

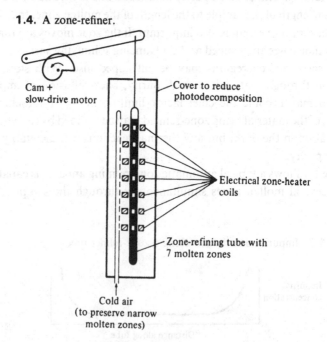

Cam +
slow-drive motor

Cover to reduce
photodecomposition

Electrical zone-heater
coils

Zone-refining tube with
7 molten zones

Cold air
(to preserve narrow
molten zones)

usually regarded as a separate stage following sample purification. Because intermolecular forces are weak, molecules arriving at a growing crystal surface are subject to only weak orientating forces. It is therefore necessary to limit the rate of growth of molecular crystals very carefully to allow time for the arriving molecules to settle to their equilibrium positions and to avoid formation of voids or defects. Optimum growth rates for molecular crystals are at least a factor of ten smaller than those for ionic crystals, where the aligning forces are strong. Crystals may be grown from vapour, solution or molten phases, each of which has its own advantages and problems.

Vapour growth

Crystals grown from the vapour-phase are solvent-free and have well-defined faces which aid orientation. They can also have low dislocation contents if grown carefully. Very thin flake crystals may be formed if vapour growth is too rapid. Slower growth, clean surfaces with a minimum of growth-nucleation sites and suitable choice of inert-gas pressure in the apparatus can help to produce large crystals by this method. For example, anthracene sublimed *in vacuo* forms bulky crystals, whereas thin flakes are formed by sublimation under 15 cmHg pressure of nitrogen[3]. Very low growth rates may be achieved by use of very small temperature gradients in the sublimation apparatus. Although this can be achieved in a cylindrical furnace with windings of increasing pitch, a more effective arrangement uses a sublimation chamber with large flat glass upper and lower faces separated by a short gap. The faces are held at temperatures high enough to sublime the sample in question, with the upper face marginally cooler than the lower face. This may be done using two electrically-heated metal blocks contacting the cell faces, with careful temperature control, or by immersing the cell in a suitable heating bath with the upper face in contact with a metal plate connected to rods protruding from the heating bath and acting as heat leaks to provide an automatic small temperature drop for the upper surface of the cell. In either case a controlled, localised additional heat leak at one point on the upper surface may temporarily be introduced to initiate formation of a seed crystallite.

Because different components generally have different vapour pressures, it is often difficult to grow uniform crystals of two-component systems (e.g. mixed crystals or charge-transfer complexes) by sublimation, but the above cell, with its short sublimation path length, has been used successfully for some two-component systems. Figure 1.5 shows an alternative arrangement[4] using a vertical furnace with a temperature gradient symmetrical about the hottest point half way up the furnace. As the sublimation tube is slowly raised through the furnace, material evaporating from the lower end condenses at the same temperature at the upper end of the tube, first in a capillary to form an oriented seed crystal, and then in the main part of the upper end of the tube. The capillary acts as a seed-selector, since crystals whose growth direction is not exactly along the capillary axis will rapidly terminate at the walls, leaving only one crystallite whose growth direction coincides with the capillary axis emerging from the capillary tip into the main part of the tube.

Melt growth

Capillary seed-selectors are also important in the Bridgmann–Stockbarger method of crystal growth from the melt[5]. A sealed ampoule of the molten sample, with a short capillary tip at its lower end, is slowly lowered through a furnace with a sharp temperature gradient between the upper zone, which is just above the melting point, and the lower zone, which is just below the melting point. The single crystallite selected by the capillary grows into a

1.5. Apparatus for crystal growth from the vapour phase.

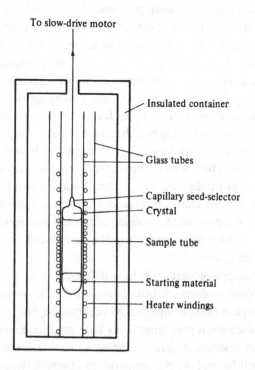

To slow-drive motor

Insulated container

Glass tubes

Capillary seed-selector
Crystal

Sample tube

Starting material

Heater windings

large single crystal as the ampoule is lowered. Disadvantages of this technique are that the crystal adopts the shape of the ampoule, leading to absence of external faces aiding orientation, and that the crystal grows at a very high temperature. Cooling to room temperature must be done carefully and slowly to minimise crystal strain, which may nevertheless still be a serious problem, particularly if the crystal sticks to the glass ampoule in places. Passing through any phase transitions which occur between the melting point and room temperature may also lead to strain. Although careful annealing can minimise these problems, the dislocation density in melt-grown crystals is still higher than in crystals grown by other techniques. Advantages of the technique are the larger size of crystals which can be grown and the fact that many impurities tend to prefer the molten phase, so that crystal growth from the melt incorporates an added stage of purification. An alternative method for growing crystals from the melt is the Kyropoulos–Czochralski method, in which a seed crystal is dipped into the surface of the molten material and withdrawn slowly so as to pull out a growing single crystal[6]. This method provides crystals with well-formed faces, less prone to strain from contact with surroundings. In practice it is difficult to use for most molecular crystals owing to the problem of sublimation of the sample material from the upper surface of the melt and onto the seed and growing crystal. Both methods are clearly unsuitable for materials which decompose on melting and for growth of mixed crystals of uniform composition; solution growth offers a better alternative for such materials.

Solution growth

Crystal growth from solution occurs well below the melting point of the solid, thus minimising risk of thermal decomposition and giving low strain and dislocation content. Also, the crystals grow freely in their natural habit, so aiding optical identification of the crystallographic axes. Incorporation of solvent into the crystals is a disadvantage which may sometimes be minimised by careful choice of solvent.

There are four main approaches to crystal growth from solution. In the simplest, the solvent is allowed to evaporate slowly from a solution at room temperature. Oxidation by the air and influx of dust particles providing an excess of nucleation sites may be avoided by allowing the solvent to evaporate in a flow of nitrogen. Careful initial filtration of the solution and use of new, unscratched flasks helps to minimise the number of nucleation sites and produce large crystals, and the flask should be left undisturbed in a fairly constant temperature environment for the duration of the crystallisation process, which may be up to several weeks.

In a second approach, a hot concentrated solution is slowly cooled in a temperature-programmed thermostat bath. As saturation is approached, a seed crystal is introduced. Stirring the solution and rotating the seed crystal slowly are both helpful in obtaining uniform crystal growth. Although this approach yields larger crystals, it is also more expensive to set up and requires a much larger amount of the material, so generally the first approach is explored initially.

The third approach uses diffusion of solvents or solutes as a means of slowly approaching the solubility limit of the material to be crystallised. For example, solutions of an electron donor and an electron acceptor may be allowed to diffuse together across porous barriers with eventual growth of crystals of charge-transfer complexes or salts, and slow diffusion of a miscible but poor solvent for a material into a solution of the material in a good solvent will gradually lower the solubility and approach saturation. This approach can work for materials which are soluble only in solvents of low volatility and which have a low temperature-coefficient of solubility, where the first two approaches would not work well.

In the fourth approach, the species forming the crystal are generated in close proximity to the growing crystal by electrochemical oxidation or reduction in a technique known as electro-crystallisation[7]. This technique is very useful for relatively rapid preparation of good-quality crystals of charge-transfer salts which form highly conducting materials[8]. Controlled current and controlled potential conditions have both been used success-fully, and the technique can be used with relatively small amounts of the materials concerned. The crystals grow from inert-metal electrodes in solution, and in some cases crystals have continued to grow even after one end of the crystal emerged above the surface of the solution, suggesting that growth actually takes place at the electrode–crystal interface.

Detailed description of the numerous variations on the methods outlined in this chapter is beyond the scope of this text. However, it is clear that the range of available methods is wide, so that with care and patience good-quality crystals of high purity should be obtainable for a great many molecular crystal materials. In view of the increasing complexity of experiments investigating properties of molecular crystals, the importance of devoting sufficient time and attention to the initial purification and crystal growth processes cannot be over-emphasised. Effort invested in these stages is nearly always amply repaid in improved experimental data and simpler data interpretation.

References

1 R.C. Sangster and J.W. Irvine, *J. Chem. Phys.*, 1956, **24**, 670.

2 W.G. Pfann, *Zone Melting*, New York: Wiley, 1966.

3 P.M. Robinson and H.G. Scott, *J. Cryst. Growth*, 1967, **1**, 187.

4 M. Radomska, R. Radomski and K. Pigon, *Mol. Cryst. Liq. Cryst.*, 1972, **18**, 75.

5 J.N. Sherwood, in *Physics and Chemistry of the Organic Solid State*, Vol. 1, ed. D. Fox, M.M. Labes and A. Weissberger, New York: Interscience, 1967.

6 J.N. Sherwood and S.J. Thomson, *J. Sci. Inst.*, 1960, **37**, 242.

7 T.C. Chiang, A.H. Reddoch and D.F. Williams, *J. Chem. Phys.*, 1971, **54**, 2051.

8 E.M. Engler, R. Greene, P. Haen, Y. Tomkiewicz, K. Mortensen and J. Berendzen, *Mol. Cryst. Liq. Cryst.*, 1982, **79**, 15.

2

Intermolecular forces

Molecular crystals differ from other classes of solids in being made up of discrete molecules. Although intramolecular forces are strong, intermolecular forces are generally weaker and short-range in their effect. This mixture of strong and weak forces is in marked contrast to the dominance of strong long-range coulombic forces in simple ionic crystals such as sodium chloride and introduces diversity to the properties of molecular crystals. Thus some properties of molecular crystals (e.g. molecular dimensions and vibrational frequencies) are essentially those of the free molecules because of the dominance of strong intramolecular forces, while others (e.g. charge transport and energy transfer) are strongly influenced by intermolecular interactions.

The structures of molecular crystals, which also affect many of the physical and chemical properties of these materials, are influenced by both intramolecular and intermolecular forces. Intramolecular forces determine molecular shapes, which in turn play an important role in determining the most effective ways of packing the molecules together in the crystal. If the intermolecular forces are particularly large or strongly dependent on the relative orientation of adjacent molecules, they may modify the crystal structure deduced from simple considerations of molecular packing. An understanding of the origins and magnitudes of intermolecular forces, and their dependence on molecular properties and intermolecular separation and orientation, is therefore essential background for understanding many properties of molecular crystals.

2.1 Interaction between dipolar molecules[1]

The electric field produced by a dipole μ along its own direction and at a distance r from its centre is $2\mu/r^3$. Hence, for two dipoles aligned head to tail at a distance r apart, the interaction energy U is given by the product of the magnitude of each dipole and the field at its centre produced by the other dipole, i.e.

$$U = -2\mu_1\mu_2/r^3. \tag{2.1}$$

In general the dipoles will not be in this ideal orientation but rather in random orientations specified by polar coordinates θ_1,ϕ_1 and θ_2,ϕ_2, respectively, as shown in figure 2.1. In considering this situation, a simple starting point is to use the components of the field of dipole 1 at dipole 2 in the z-direction (F_{par}) and perpendicular to the z-direction in a plane containing the direction of μ_1 and the z-axis (F_{perp}), which are given by[2]

$$F_{par} = 2\mu_1\cos\theta_1/r^3 \tag{2.2}$$
and $\quad F_{perp} = \mu_1\sin\theta_1/r^3. \tag{2.3}$

Now, the component of μ_2 in the z-direction is $\mu_2\cos\theta_2$, so there is an attractive force

$$U_{att} = -F_{par}\mu_2\cos\theta_2 = -2\mu_1\mu_2\cos\theta_1\cos\theta_2/r^3. \tag{2.4}$$

However, both dipoles have parallel components which lie perpendicular to the z-direction and in the plane defined above, and these interact repulsively. The component of μ_2 in this direction is $\mu_2\sin\theta_2\cos(\phi_2-\phi_1)$, so the repulsive force is

$$U_{rep} = (\mu_1\mu_2/r^3)\sin\theta_1\sin\theta_2\cos(\phi_2-\phi_1) \tag{2.5}$$

and the net interaction is given by

$$U = -(\mu_1\mu_2/r^3)\{2\cos\theta_1\cos\theta_2 - \sin\theta_1\sin\theta_2\cos(\phi_1-\phi_2)\}. \tag{2.6}$$

Hence, for two dipolar molecules held in fixed relative orientations in a crystal, the dipole–dipole interaction energy is proportional to r^{-3}. It is

2.1.

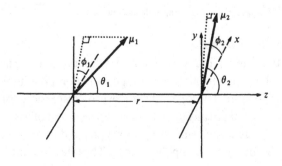

important to notice that this r^{-3}-dependence is different from the r^{-6}-dependence of the interaction energy of two dipolar species in solution. In the latter case, if all orientations of the two dipoles were equally probable, the time-averaged interaction energy would be zero. Application of Boltzmann statistics, to allow for the fact that orientations producing lower interaction energy are more probable than those giving high interaction energy, leads to an average interaction energy $\bar{U}$ given by

$$\bar{U} = -2\mu_1^2\mu_2^2/3r^6kT \tag{2.7}$$

where T = temperature in K.

This expression is not generally applicable to molecular crystals, as the molecules are not usually free to rotate in all directions and the statistical approach is no longer valid.

2.2   Dipole-induced dipole interactions[3]

The electric field $(F = 2\mu_1/r^3)$ of one dipole, whose component along the line joining it to a second polarisable molecule centred at a distance r away is μ_1, induces a dipole on this second molecule given by

$$\mu_{ind} = -\alpha_2 F = -2\alpha_2\mu_1/r^3 \tag{2.8}$$

where α_2 is the molecular polarisability.

The interaction energy of the permanent dipole with this induced dipole is thus

$$U = F\mu_{ind} = -4\alpha_2\mu_1^2/r^6. \tag{2.9}$$

This interaction is always attractive, irrespective of the relative orientation of the two molecules, although its magnitude may depend on this orientation if the molecule being polarised has an anisotropic polarisability. Since polar molecules may also be polarised, this interaction also contributes to the total interaction energy of two dipolar molecules, i.e.

$$2U_{total} = U_{dipole-dipole} + \sum U_{dipole-induced\ dipole}. \tag{2.10}$$

2.3   Dispersion forces[4]

Interactions involving permanent dipoles do not explain the cohesive forces holding together crystals composed of non-polar molecules, e.g. anthracene. London developed the theory of dispersion forces as follows. Consider two spherically symmetrical species with polarisability α, and make small charge displacements r_1 $(=x_1y_1z_1)$ and r_2 $(=x_2y_2z_2)$ on each, leading to the formation of dipolar species. The total potential energy is then the sum of the energy required to produce the charge displacements (I) and the energy of interaction of the two resulting dipoles (II):

$$V = e^2 r_1^2/2\alpha + e^2 r_2^2/2\alpha + (e^2/r^3)(x_1 x_2 + y_1 y_2 - 2z_1 z_2).$$
$$\text{I} \qquad\qquad\qquad\qquad \text{II} \qquad\qquad\qquad (2.11)$$

(Term II is the same as equation 2.6, allowing for the different notation required here.)

If the displacements are now re-defined in terms of normalised coordinates $(x_+ = (x_1 + x_2)/\sqrt{2}; \ x_- = (x_1 - x_2)/\sqrt{2}; \ y_+ = (y_1 + y_2)/\sqrt{2}$, etc.), the potential-energy expression becomes

$$V = (e^2/2\alpha)[(1 + \alpha r^{-3})(x_+^2 + y_+^2) + (1 - \alpha r^{-3})(x_-^2 + y_-^2) + (1 - 2\alpha r^{-3})z_+^2 + (1 + 2\alpha r^{-3})z_-^2] \qquad (2.12)$$

which is a sum of squares representing the potential energy of six independent oscillators, with frequencies $v_{x+}, v_{y+}, v_{z+}, v_{x-}, v_{y-}, v_{z-}$ given by

$$v_{x+} = v_{y+} = v_0 \sqrt{(1 \pm \alpha r^{-3})} \approx v_0 \{1 \pm (\alpha/2r^3) - (\alpha^2/8r^6) \pm \ldots\}$$
$$v_{z+} = v_0 \sqrt{(1 \mp 2\alpha r^{-3})} \approx v_0 \{1 \mp (\alpha/r^3) - (\alpha^2/2r^6) \mp \ldots\} \qquad (2.13)$$
$$\text{where } \alpha \ll r^3$$
$$v_0 = e/(\sqrt{m\alpha}) \text{ and } m = \text{reduced mass of each species.}$$

The lowest energy of this system of six oscillators is given by

$$E_0 = (h/2)(v_{x+} + v_{y+} + v_{z+} + v_{x-} + v_{y-} + v_{z-})$$
$$= (hv_0/2)[6 + (\tfrac{1}{2} + \tfrac{1}{2} - 1 - \tfrac{1}{2} - \tfrac{1}{2} + 1)(\alpha/r^3) - \{(4/8) + (2/2)\}(\alpha^2/r^6) + \ldots]$$
$$= 3hv_0 - (3/4)hv_0 \alpha^2 r^{-6} + \ldots. \qquad (2.14)$$

The term $3hv_0$ is the zero-point energy of the two isolated species, while the second term $-(3/4)hv_0 \alpha^2 r^{-6}$ is the attractive dispersion energy. The frequency v_0 is a characteristic frequency for one of the isolated species and also appears as the frequency at which refractive dispersion increases rapidly (e.g. when hv_0 is equal to the energy of a strongly allowed electronic transition or the ionisation energy of the species), which is why the name 'dispersion force' is used.

In practice, more complex charge displacements giving rise to quadrupole and higher multipoles are strictly required for a full description of the state of a molecule, so that the dispersion force should be written as

$$U_{\text{dispersion}} = c_6 r^{-6} + c_8 r^{-8} + c_{10} r^{-10} + \ldots \qquad (2.15)$$

where c_6, c_8, etc. are constants.

Quadrupole and higher-order terms are often neglected but can be very important, for example the heat of sublimation of crystalline CO_2 at 0 K is 27 kJ mol^{-1}, of which 45% is the result of quadrupole–quadrupole interactions and 55% of dispersion forces. For more detailed consideration of these and other effects, the reader is referred to reference 5. For all except the smallest molecules (such as CO_2 above), the actual distribution of charge over the array of atoms constituting the molecule does not correspond physically with either a point dipole or a point quadrupole, and the method of atom–atom potentials, to be described later in this chapter, provides a more realistic and useful description.

Dipole–dipole, dipole–induced dipole and dispersion forces are often referred to collectively as van der Waals forces. The expressions for their magnitudes discussed above show that van der Waals forces should be largest between polar or polarisable molecules. In fact, contrary to common intuitive impressions, dispersion forces can often be comparable in magnitude with dipole–dipole interactions. For example, the sublimation energy of argon at 0 K is as high as 40% of that of the isoelectronic polar molecule HCl. The melting point data in table 2.1 also reflect trends in van der Waals forces. Comparison of *o*-xylene and cyclooctatetraene, both containing eight carbon atoms, shows that cyclooctatetraene, which has more polarisable π-electrons, has the higher melting point. Similarly, benzene has a higher melting point than *n*-hexane. In the series of aromatics from benzene to hexacene, as the size of the π-electron system increases the polarisability increases and hence the melting point also increases. (The low melting point of pentacene, or alternatively the high melting point of tetracene, is an intriguing anomaly in this series!) All these comparisons are for hydrocarbons. When molecules of widely different composition are compared, it is important to remember that the r^{-6} term will vary as well as

Table 2.1 *Melting points (°C) of some molecular crystals*

o-Xylene	−25
Cyclooctatetraene	−4
n-Hexane	−95
Benzene	5.5
Naphthalene	80
Anthracene	216
Tetracene	357
Pentacene	270
Hexacene	380

the molecular polarity or polarisability. Thus, for example, the size difference between fluorine and hydrogen dominates over effects resulting from differing polarisability, and fluorocarbons are more volatile than hydrocarbons.

2.4 Charge-transfer interactions[6]

Interactions between an electron donor molecule (D) and an electron acceptor molecule (A) are stronger than van der Waals forces and lead to charge-transfer complex formation. Mulliken explained these charge-transfer resonance forces in terms of a new ground-state wave-function (Ψ_G) in which charge-transfer states ($\Psi_{D^+A^-}$) are mixed with the normal ground-state wave-function of the complex (Ψ_{DA}). Ψ_{DA} has energy (W_0) equal to the sum of the separate energies of D and A together with all other contributions to their interaction except the charge-transfer contribution (e.g. van der Waals forces, hydrogen bonding, etc.). In the simplest case, where only one charge-transfer state is considered,

$$\Psi_G = a\Psi_{DA} + b\Psi_{D^+A^-} (a \gg b). \tag{2.16}$$

Provided the charge-transfer interaction is weak (as implied by $a \gg b$), second-order perturbation theory can be used to show that the energy (W_G) of the new ground state (Ψ_G) is given by

$$W_G = W_0 - (H_{01} - SW_0)^2/(W_1 - W_0) \tag{2.17}$$

where W_1 is the energy of the state whose wave-function is $\Psi_{D^+A^-}$

the Hamiltonian $H_{01} = \int \Psi_{DA} H \Psi_{D^+A^-} \, dv$ (2.18)

and the overlap integral $S = \int \Psi_{DA} \Psi_{D^+A^-} \, dv.$ (2.19)

The importance of this expression is that the charge-transfer contribution to the interaction will be large if the final term $\{(H_{01} - SW_0)^2/(W_1 - W_0)\}$ is large. This implies that $W_1 - W_0$ should be small, i.e. that the energy to transfer an electron from D to A should be small, which will be true if D has a low ionisation potential and A has a high electron affinity. It also implies that $(H_{01} - SW_0)^2$ should be large for large charge-transfer stabilisation. This means that the overlap of Ψ_{DA} and $\Psi_{D^+A^-}$ should be large, i.e. the orbital donating the electron and the orbital receiving the electron should be close enough physically to overlap significantly, and ideally should have the same symmetry. Thus charge-transfer interactions are strongest between donor orbitals of high energy and acceptor orbitals of low energy, provided they have similar symmetry and are physically able to approach closely.

The symmetry requirement means that charge-transfer contributions to intermolecular forces are strongly dependent on the relative orientation of the donor and acceptor molecules. This can be illustrated by considering the interactions between the two highest occupied orbitals of naphthalene and the lowest vacant orbital of tetracyanoethylene (TCNE) (figure 2.2). Examination of these orbital symmetries shows that the orientation of figure 2.3 gives good symmetry matching between the naphthalene HOMO and the TCNE LUMO. This orientation, however, gives zero overlap between the second HOMO of naphthalene and the TCNE LUMO (the nodal planes, denoted by broken lines, are perpendicular). Conversely, the orientation maximising overlap between the second HOMO of naphthalene and the TCNE LUMO is that shown in figure 2.4, and this in turn gives zero overlap between the naphthalene HOMO and the TCNE LUMO.

In general, charge-transfer interactions occur between all available donor orbitals (each with its own energy and symmetry) and all available acceptor orbitals, so equation 2.16 should be rewritten as

$$\Psi_G = a\Psi_{DA} + \sum_i b_i \Psi_{iD^+A^-} \qquad (2.20)$$

with the summation over all i possible combinations of donor and acceptor orbitals. The total stabilisation resulting from charge-transfer interaction is then the sum of the individual contributions from all possible combinations of donor and acceptor orbitals, calculated by expressions analogous to equation 2.17. Molecular orbital methods have been developed for performing such calculations and provide estimates of the total magnitude of the charge-transfer interactions for given donor–acceptor orientations, as well as prediction of the orientation maximising these interactions and of the sensitivity of the interaction energy to changes in orientation[7].

2.2

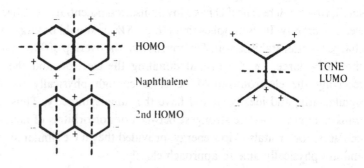

2.5 Repulsive forces

The closest distance of approach of two molecules is determined by the point at which the attractive forces are exactly balanced by the repulsive forces which arise when the electron clouds of two molecules begin to penetrate each other significantly. These repulsive forces have two main origins. It follows from the Pauli principle that two electrons can occupy the same volume element of space only if they have sufficiently different velocities. This implies that energy must be given to electrons in the region of interpenetration, which is one origin of repulsive forces. Secondly, since electrons will therefore tend to avoid the region of interpenetration, they no longer screen the nuclear charges on the molecules so effectively and coulombic repulsion between nuclei on the two molecules increases.

Ab initio calculations of these repulsive forces for particular molecules is complex, requiring, for example, detailed knowledge of the charge distribution on the periphery of the molecule as a function of the approach distance of the second molecule. A more practical approach is to develop empirical repulsive potentials which can be fitted to experimental data (e.g. compressibility data). The two most commonly used repulsive potentials have the forms

$$ar^{-n} \text{ (with } n \text{ commonly 12) and } be^{-cr},$$

where a, b, c and n are empirical constants.

The former expression is widely used and has the advantage of simplicity, although the exponential expression is considered ultimately more realistic. Both these expressions assume isotropic repulsions between atoms, whereas chemists have long realised that in many cases lone-pairs of electrons, d-orbitals, etc. lead to pronounced anisotropy in atomic interactions. Recently this has been incorporated into anisotropic repulsion potentials, which (as will be shown in chapter 3) are required in order satisfactorily to model crystal structures ranging from that of a molecule as simple as chlorine[8] to those of the azabenzene series[9].

2.3   **2.4.**

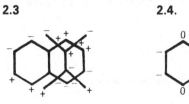

2.6   Atom–atom potentials[10]

Although the foregoing discussion is useful on a qualitative level for predicting molecular properties leading to strong intermolecular forces and the structural implications of the different intermolecular forces, it is less useful quantitatively. It is also unrealistic, as has already been mentioned, to discuss interactions between two molecules of finite size and possibly complex shape in terms of interactions between equivalent point dipoles, quadrupoles, etc. A more realistic and quantitative approach is to sum the interactions of each of the atoms of a molecule at some convenient origin with atoms of surrounding molecules. Since most intermolecular forces fall off rapidly with increasing distance (e.g. r^{-6}), it is normally not necessary to extend the summation to very large distances. A typical approach is to partition the interactions between atoms of types k and l at a distance r_{ij} into electrostatic, repulsive and van der Waals potential energies, giving an expression for the interaction energy $U^{kl}(r_{ij})$ of the form

$$U^{kl}(r_{ij}) = (q_i q_j)/(Dr_{ij}) + A^{kl}/r_{ij}^{12} - c^{kl}/r_{ij}^6 \qquad (2.21)$$

where q_i and q_j are the fractional charges on the atoms,
        D is the effective dielectric constant,
        A^{kl} is the repulsive coefficient,
and c^{kl} is the attractive coefficient.

These constants can be evaluated by optimising the fit of the calculated lattice energy to the experimental values for several model compounds, with distinct values for the interactions of C–C, C–H, H–H, etc. These methods have been successfully used to estimate lattice energies, predict structures of molecular crystals, estimate the relative stability of proposed new structural modifications or defects and predict barriers to molecular motion in crystals. It is likely that these calculations will play an increasingly important role in the study of molecular crystals as computing power continues to grow, particularly with the advent of distributed-array processors, although the traditional approach to intermolecular forces can still be a useful predictive guide to chemists and materials scientists in search of new materials.

References

1 W.H. Keesom, *Proc. K. Akad. Wetenschappen Amsterdam*, 1912, **15**, 417.
2 P. Debye, *Physik. Z.*, 1920, **21**, 178.
3 W.J. Duffin, *Electricity and Magnetism*, London: McGraw-Hill, 1973, p. 79.
4 F. London, *Trans. Faraday Soc.*, 1937, **33**, 8.

5 A.D. Buckingham, in *Physics of Dielectric Solids*, ed. C.H.L. Goodman, Institute of Physics Conference Series No. 58, 1980, p. 113.
6 R.S. Mulliken and W.B. Person, *Molecular Complexes. A Lecture and Reprint Volume*, New York: Wiley, 1969.
7 C.K. Prout and B. Mayoh, *J. Chem. Soc. Faraday II*, 1972, **68**, 1072.
8 S.L. Price and A.J. Stone, *Mol. Phys.*, 1982, **47**, 1457.
9 S.L. Price and A.J. Stone, *Mol. Phys.*, 1984, **51**, 569.
10 D.E. Williams, *Acta Cryst.*, 1972, **A28**, 629.

3

Crystal structures

The structures of molecular crystals are conveniently classified into three groups depending on the number of different intermolecular forces operating. The simplest group is composed of non-polar molecules where only dispersion and repulsion forces operate. The second group is composed of molecules with polar substituents where dipole–dipole and dipole–induced dipole forces also operate. Hydrogen-bonded molecular crystals form a special case within this group. Finally, in the third group are crystals in which charge-transfer interactions occur between different molecules forming electron donor–acceptor complexes or charge-transfer salts.

3.1   Crystals of non-polar molecules

For the first group, with relatively weak non-directional dispersion forces and strong short-range repulsive forces, a useful starting point is the idea that molecules will tend to attain an effective packing arrangement which minimises intermolecular repulsions. This idea has been extensively developed by Kitaigorodskii[1]. A parameter for judging the effectiveness of molecular packing in a crystal is the packing coefficient K, defined as

$$K = ZV_0/V \tag{3.1}$$

where V is the volume of a crystallographic unit-cell containing Z molecules, each molecule having a volume V_0 calculable from knowledge of atomic radii and molecular dimensions.

Table 3.1 lists some packing coefficients for aromatic hydrocarbons. Although K increases with the size of the aromatic molecule and depends on molecular shape (e.g. compare anthracene 0.722 and phenanthrene 0.684), the most striking feature of this table is the small difference between

K for different molecules. This strongly suggests that the lattice packing is heavily influenced by close-packing considerations. In fact, calculations of the structures of crystals of such molecules using atom–atom potentials based only on intermolecular repulsive forces and dispersion forces often yield results close to the observed structures in overall molecular arrangement, if not in precise lattice constants[2]. In these calculations the objective is to calculate the total energy of interaction between molecules in a lattice volume as large as permitted by computer storage and time constraints. The structure which minimises the total interaction energy is then compared with the observed structure. Although this approach is useful in confirming the importance of good molecular packing, it is not generally necessary as a means of determining structures of molecular crystals, because accurate crystal structure determinations by diffraction methods are now available for a vast range of these materials.

The requirement of efficient lattice packing leads in practice to certain characteristic lattice arrangements which occur frequently in the structures of crystals of this first class of molecule. For example, many large planar molecules form structures composed of molecules packed face-to-face in stacks. Three possible arrangements of such stacks are shown in figure 3.1. In practice the simplest arrangement (*a*) is less common than (*b*) or (*c*) in homomolecular crystals. This is because even planar molecules have surfaces that are not flat, but rather are 'knobbly', for they are assemblies of essentially spherical atoms. Stacking arrangement (*a*) places atoms of one molecule directly above or below corresponding atoms in the neighbouring molecule, an arrangement which produces greater atom–atom repulsion than (*b*) or (*c*), in which the inclination of the molecules to the stacking axis

Table 3.1 *Packing coefficients* (K) *of aromatic hydrocarbons*

	K
Benzene	0.681
Naphthalene	0.702
Anthracene	0.722
Chrysene	0.737
Perylene	0.805
Coronene	0.726
Graphite	0.887
Phenanthrene	0.684
cf. spheres in	
fcc packing	0.741

3.1. Stacking arrangements of planar molecules in crystals. The examples are (*a*) anthracene/tetracyanoquinodimethane (TCNQ)[3], (*b*) hexamethylbenzene[4] and (*c*) naphthalene[5].

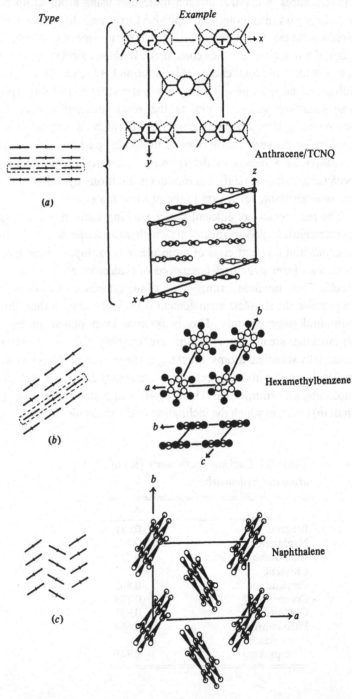

displaces corresponding atoms of adjacent molecules laterally. (Note however that these repulsions in arrangement (*a*) can be avoided if a screw axis of symmetry runs along the stack axis (c.f. figure 3.9, below p. 30).) As well as achieving effective molecular packing, these stacking arrangements maximise dispersion forces between planar molecules having delocalised π-electron systems, as the polarisable π-electron clouds of adjacent molecules are close to each other.

Structures containing stacks composed of molecules of one type are termed *homosoric* (Greek *sorus* = pile or stack). It is worth noting at this point that *heterosoric* structures, with stacks of alternating molecules of two types, can adopt stacking arrangement (*a*) more easily, since molecules above and below any given molecule in the stack have different structures. Arrangement (*a*) does not necessarily place atoms of one molecule directly above or below atoms of adjacent molecules in this case and atom–atom repulsions may thus be minimised even in this arrangement. Electron donor–acceptor complexes form the largest class of heterosoric molecular crystals.

From an alternative point of view, stacking arrangements (*a*) and (*b*) can be regarded as structures built up of successive layers of close packed molecules. The broken lines on figure 3.1(*a*) and (*b*) outline three molecules in one such layer. The difference between figure 3.1(*a*) and (*b*) is then that successive layers of molecules are displaced laterally in (*b*) but not in (*a*). Unfortunately, the diagrams most commonly used in the crystallographic literature to show the arrangements of molecules in the unit cell are projections along one of the crystallographic axes, and these do not always reveal the beauty and efficiency of packing of molecules in layers in the structures. A more effective projection for this purpose is that onto the least-squares best plane through one of the molecules. Figures 3.2 and 3.3

3.2. Perylene/fluoranil (stack)[6].

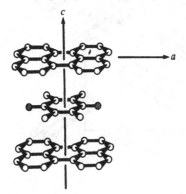

show the structure of the perylene/fluoranil complex[6] projected in these two ways. The former projection emphasises that the structure is composed of stacks of molecules, while the latter emphasises the close-packed sheets formed by molecules in adjacent stacks.

A variation on these stacking arrangements that is occasionally found involves pairs of molecules stacked together in the herringbone fashion of figure 3.1(c) as shown in figure 3.4. Pyrene is one example of this lattice arrangement[7], while perylene, interestingly, crystallises in two forms, α-perylene having the pairwise packing of figure 3.4 and β-perylene having the herringbone packing of figure 3.1(c)[8]. These two forms of perylene must

3.3. Perylene/fluoranil (sheet)[6].

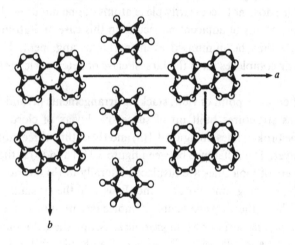

3.4. Pyrene[7].

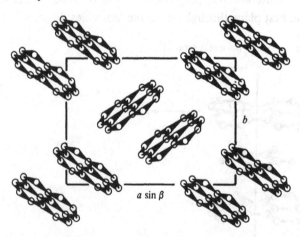

have very similar lattice energies, for crystallisation from solutions of perylene produces a mixture of crystals of the two forms, but the crystal habits are different, and the different molecular packing profoundly modifies the physical properties, as will be discussed in later chapters. The densities of the two forms differ by less than 5%, implying similar packing coefficients.

The existence of phases of different lattice structure but similar lattice energy has important consequences for the physical properties of molecular crystals. Thus, phase transitions are common; for example, hexamethyl-benzene changes from a triclinic structure to a monoclinic structure at 110°C, with minor displacements of the molecules and an associated halving of the activation energy for in-plane rotation of the molecules on their lattice sites[4]. Twinning, in which changes in stacking arrangement lead to crystals composed of two or more regions of similar structure but different orientation, as shown in figure 3.5 for example, is also a common problem. Sometimes even lattice deformations induced by physically straining a crystal may lead to the formation of very small regions having new lattice structures occurring as islands within a crystal whose majority structure appears unchanged. The lattice constants of these small regions have been deduced by advanced electron-microscopic techniques and their structures deduced from atom–atom potential calculations. For example, straining anthracene crystals introduces regions of triclinic structure which may have important consequences for solid-state reactions such as photodimerisation[9]. These phenomena will be discussed in later chapters.

Another characteristic lattice packing arrangement occurs for molecules of this first group, which are either of tetrahedral symmetry or are effectively spherical or ellipsoidal[10]. At temperatures approaching the

3.5.

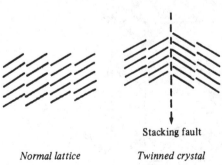

Stacking fault

Normal lattice   *Twinned crystal*

melting point, crystals of these molecules are soft and plastic, and the molecules are orientationally disordered on their lattice sites. One of the most common structures for such plastic crystals is face-centred cubic. This is the lattice adopted by neon, argon, krypton and xenon and reflects efficient packing of almost-spherical molecules. Examples of molecules whose crystals exhibit such plastic phases with face-centred cubic structure include methane, 2,2-dimethylpropane, 2,2-dichloropropane, carbon tet-rachloride, cyclohexane and adamantane (figure 3.6).

Despite the success of simple packing considerations in accounting for the observed structures of the crystals of many of these non-polar molecules, there are several quite simple molecules whose observed crystal structures cannot be explained in terms of traditional atom–atom potentials, even with the inclusion of quadrupole–quadrupole interactions. For example, chlor-ine crystallises in the structure shown in figure 3.7, with the molecules lying in the layer planes of the lattice in the C_{mca} space-group[12]. Accurate X-ray diffraction data reveal a torus of lone-pair electron density surrounding each chlorine atom in the positions indicated in figure 3.7. This may well arise from two directed lone-pairs on each chlorine atom averaged by rotation around the Cl–Cl bond axis. The crystal structure minimises repulsions by placing lone-pairs of one molecule neatly between lone-pairs of a neighbour, but is not that expected from isotropic atom–atom potential calcula-tions[13-15]. However, it has been successfully modelled using anisotropic repulsion parameters in the atom–atom potential calculation[16].

Similarly, this approach has been successful in accounting for the very

3.6. Adamantane[11].

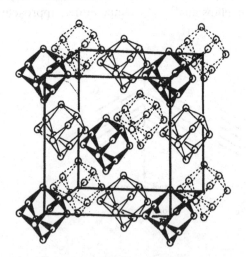

diverse crystal structures of the azabenzene series of molecules. Although these are all six-membered ring molecules, s-tetrazine[17] has a herringbone lattice (figure 3.8, c.f. figure 3.1(c)), s-triazine[18] has stacks of parallel molecules (figure 3.9, c.f. figure 3.1(a)) related by a three-fold screw axis, pyrazine[19] and pyrimidine[20] are again herringbone type while pyridine[21] crystallises in a very complex structure with 16 molecules/cell comprising four crystallographically-independent molecules each with 14 nearest neighbours (figure 3.10). In this series of molecules, the different electro-negativities of carbon and nitrogen atoms lead to uneven charge

3.7. The crystal structure of chlorine[12]. (Small circles denote chlorine atoms, with the closed circles 3.104 Å (a/2) below the open circles. Large circles represent the cross-section of a torus of electron density corresponding to each lone-pair of electrons.)

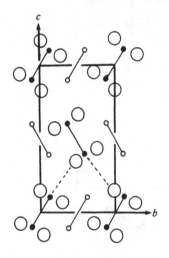

3.8. s-Tetrazine[17].

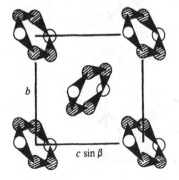

distributions in the rings, and hence electrostatic interactions between molecules must be incorporated in addition to anisotropic repulsion parameters to account for the observed structures[22]. Such molecules thus form a link with the second category of crystal structures.

3.9. *s*-Triazine[18].

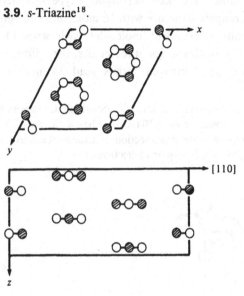

3.10. Pyridine[21].

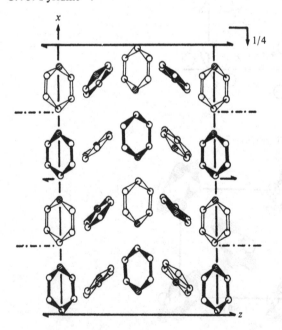

3.2   Crystals of molecules having polar substituents

As seen in chapter 2, the interactions between two dipoles of fixed orientation vary as r^{-3} in contrast to the r^{-6}-dependence of induction and dispersion forces. It might therefore be expected that structures of this group would reflect the orientations maximising dipole–dipole or dipole–induced dipole interactions. There are several examples of structures showing these orientations. Figure 3.11 shows the structure of *p*-iodobenzonitrile[23] in which dipolar interactions lead to the formation of chains of molecules with strong attraction between the CN end of one molecule and the I end of the next (N .... I distance = 3.18 Å, c.f. van der Waals contact distance = 3.65 Å). The chains pack efficiently into layers, with the molecular dipoles all in the same direction in a given layer and in opposite directions in successive layers.

Hydrogen-bonded molecular crystals frequently have rather open structures which optimise the dipolar hydrogen-bonding interactions but do not give efficient molecular packing (for example, the packing coefficient of ice is only 0.38). Figure 3.12 shows the structure of quinol as a typical example[24]. This structure contains quite large holes in the open three-dimensional network of hydrogen-bonded molecules (whose packing coefficient is 0.64, marginally lower than for benzene), and it is possible to fit a variety of foreign molecules into these holes, forming *clathrate* structures. Thus, hydroquinone forms clathrates of the type $3[C_6H_4(OH)_2]X$, where X can be SO_2, HCl, HBr, H_2S, ethyne, methanol, acetonitrile,

3.11. *p*-Iodobenzonitrile[23].

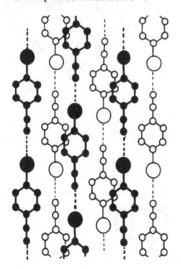

methanoic acid and CO_2. Once these molecules are trapped in the crystal they are not easily lost, although the interactions between them and the hydroquinone molecules are weak, so that if the hole is larger than strictly necessary to accommodate the molecule, it may spin around inside the hole.

Benzoquinone and naphthoquinone[25] also adopt structures which place a polar carbonyl group of one molecule above the centre of the polarisable π-electron system of the six-membered ring of the next, favouring dipole–induced dipole interactions (figure 3.13). However, these structures in which successive molecules in a stack are displaced laterally also reduce atom–atom repulsions as already discussed, and their occurrence does not prove that dipole interactions are the dominant structure-determining feature. Thus, in 1,4-anthraquinone (figure 3.14), the carbonyl groups no longer lie over the centres of the six-membered rings and the relative orientation of successive molecules seems to be dictated more by the need to stagger atoms in successive layers than by dipolar interactions.

Further evidence showing that even the structures of molecules with highly polar substituents may be more influenced by packing than by electrostatic interactions is found in the structures of TCNE[26] and

3.12. Quinol[24].

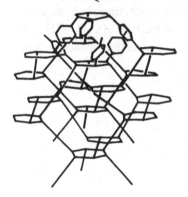 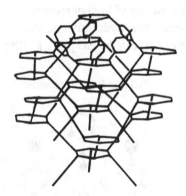

3.13. Benzoquinone[25]. Naphthoquinone[25].

tetracyanoquinodimethane (TCNQ)[27]. In the cubic form of TCNE, molecules are packed very effectively in a way which is not especially favourable for optimising electrostatic interactions (figure 3.15). In the case of TCNQ, the correct crystal structure was selected from a series of trial models simply on the grounds of optimising molecular packing in the known unit cell, without invoking any electrostatic interactions.

In general, therefore, dipolar interactions rarely lead to structures which are in conflict with the requirements of efficient lattice packing, except in the case of strongly hydrogen-bonded crystals, although some minor modifications of normal close-packed structures may occur (e.g. *p*-iodobenzonitrile). The converse, namely that the lattice structures of all molecules with dipolar substituents can be predicted from packing considerations alone, does not follow. As already noted, there are many cases where a molecule can form several crystalline phases of different structure (polymorphs) but similar lattice energy, so that dipolar interactions can lead to the adoption of a structure which is slightly less favourable than the ideal on packing considerations. This situation has a parallel in ionic crystals, where the fine balance in the electrostatic energy of sodium chloride and caesium chloride structures may be tipped one way or the other according to the magnitude of dispersion forces etc.

3.14. 1,4-Anthraquinone[25].

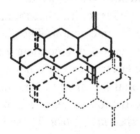

3.15. Tetracyanoethene[26].

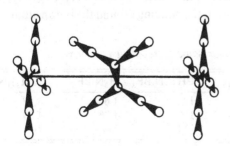

3.3   Structures involving charge transfer

In the final group of structures, electron donor–acceptor interactions occur which can affect inter- and intramolecular distances as well as the relative orientations of donor and acceptor molecules in the crystal. Thus, these interactions increase intermolecular attractive forces and may lead to shorter donor–acceptor distances than expected from consideration of van der Waals radii, particularly if the donor–acceptor interaction is localised. At the same time the partial transfer of electrons from one molecule to another may lead to changes in intramolecular bond lengths and angles. These changes will also be most marked if the electron is transferred to or from particularly localised orbitals. As shown in chapter 2, the additional stabilisation from charge-transfer resonance depends on overlap of donor and acceptor orbitals, so the strength and structural influence of charge-transfer interactions is optimised when the relative orientation of donor and acceptor molecules is such as to maximise this overlap. This is Mulliken's *Overlap and Orientation principle*. In practice, any of these structural influences of donor–acceptor interactions may be obscured if they conflict with the structural requirements imposed by other intermolecular forces, for example packing requirements or strong dipole interactions. Nevertheless, there are many examples of crystalline donor–acceptor complexes where such conflicts do not occur and where the structural effects of charge-transfer interactions are observed.

Mulliken classified donor–acceptor complexes according to the nature of the orbitals involved in the charge-transfer interaction. Thus, electron donation can be from a non-bonding lone-pair (n), or a bonding orbital (σ or π), while the orbital which accepts the electron can be a non-bonding vacant orbital (v) or an anti-bonding orbital (σ^* or π^*). Donation from n-type orbitals to v-type orbitals leads to the formation of a strong new bond, and such complexes are therefore called *increvalent*. Table 3.2 gives data for increvalent complexes involving donation from a nitrogen-atom lone-pair to the vacant orbital of trifluoroborane[28]. As the lone-pair ionisation energy decreases in the order $CH_3CN > H_3N > (CH_3)_3N$, the N ... B distance decreases and the bonding around the boron atom changes

Table 3.2 *Structural data for some increvalent complexes*

	BF_3	$CH_3CN.BF_3$	$H_3N.BF_3$	$(CH_3)_3N.BF_3$
N–B distance (Å)	–	1.63	1.60	1.58
B–F distance (Å)	1.30	1.33	1.38	1.39
F–B–F angle (°)	120	114	111	107

from trigonal planar sp^2 to tetrahedral sp^3 with a consequent parallel increase in the B–F bond length.

Electron donation from σ or π bonding orbitals and electron acceptance by antibonding σ^* or π^* orbitals leads to weakening of the intramolecular bonding, and hence the term *sacrificial* donor or acceptor, respectively. Table 3.3 contains structural data for some n–σ^* sacrificial complexes involving the iodine molecule as electron acceptor. As the degree of charge transfer increases going down the table, the I–I bond length increases, caused by the larger degree of electron donation to σ^* orbitals (bond length in free $I_2 = 2.67$ Å). Also, the donor–acceptor intermolecular contact distance, which is shorter than the normal van der Waals contact distance but longer than the sum of the covalent radii of the contacting atoms, decreases within groups of complexes involving the same donor atom, reflecting the increasing contribution of charge-transfer interactions to the intermolecular forces. The localised charge-transfer interactions in n–v and n–σ^* complexes also provide clear examples of the influence of Mulliken's Overlap and Orientation principle. Figure 3.16 shows the dithian/2SbI$_3$ complex structure. The sulphur lone-pairs point along the ideal axial and equatorial directions, while lattice packing requirements lead to S–S–Sb angles differing from the ideal. The overlap between the lone-pair and vacant orbital on Sb will be largest when the S–S–Sb angle is closest to ideal, and the observed S–Sb distance is therefore shorter for the equatorial contact than the axial contact, as predicted by the Overlap and Orientation principle.

The π–π^* charge-transfer interactions are more delocalised, and their influence on inter- and intramolecular distances is correspondingly less marked. Molecular complexes between π-donors and π-acceptors in which

Table 3.3 *Structural data for some sacrificial complexes*

Donor	I–I distance (Å)	D–A contact	Observed distance (Å)	Van der Waals radius sum (Å)	Covalent radius sum (Å)
1,4 Dithiane	2.787	S–I	2.867	4.00	2.37
Benzyl sulphide	2.82	S–I	2.78	4.00	2.37
4.Picoline	2.83	N–I	2.31	3.65	2.03
Trimethylamine	2.83	N–I	2.27	3.65	2.03
1,4 Diselenane	2.870	Se–I	2.829	4.15	2.50
Tetrahydroselenophene	2.914	Se–I	2.762	4.15	2.50
1-Oxa-4-selenacyclohexane	2.956	Se–I	2.755	4.15	2.50

the degree of charge transfer in the ground state is small crystallise in the lattice types of figure 3.1 with heterosoric stacks composed of alternating donor and acceptor molecules. In the great majority of cases, the bond lengths and angles of the donor and acceptor molecules in the complex are identical to those found in homomolecular crystals of the donor and acceptor, within the limits of accuracy of the structure determinations. The donor and acceptor molecules in the complexes are usually close to planar, although systematic minor deviations from planarity are common, particularly in cases where the electron acceptor does not lie symmetrically above or below the centre of the electron donor. In such cases also the planes of donor and acceptor may be tipped out of parallel by up to 18° by lattice packing effects.

Lattice packing, particularly repulsion between contacting atoms of neighbouring donors and acceptors within a stack, also influences the mean separation of the molecular planes and obscures any general correlation with the strength of the charge-transfer interaction. Indeed, as pointed out in chapter 2, the charge-transfer stabilisation does not depend simply on donor ionisation potential and acceptor electron affinity, but rather on the sum of the interactions between all donor filled-orbitals and all acceptor vacant-orbitals (c.f. equations 2.16, 2.17 and 2.20). Evaluation of this sum involves knowledge of all the relevant orbital energies (not simply ionisation potential and electron affinity, which refer, respectively, to the donor highest occupied molecular orbital and the acceptor lowest unoccupied molecular orbital) as well as the relative orientation of donor and acceptor, which influences the overlap of interacting orbitals. A correlation

3.16. 1,4-Dithian/2SbI$_3$[29].

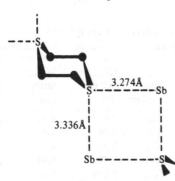

		Equatorial	Axial
S–S–Sb angle:	observed	147.3	124.6
	ideal	146	105

between donor–acceptor contact distance and donor ionisation potential or acceptor electron affinity is thus expected in a series of complexes only if:

(1) intermolecular repulsion is similar across the series;
(2) interactions between the donor HOMO and acceptor LUMO either dominate interactions of other donor and acceptor orbitals or are paralleled in magnitude by these other interactions;
(3) relative orientations of donor and acceptor are equally favourable to the dominant charge-transfer interaction across the series.

These conditions are fairly well fulfilled in n–v or n–σ^* complexes, but much more difficult to fulfil in π–π^* complexes. The latter frequently have several high-lying occupied donor orbitals or low-lying vacant acceptor orbitals of different symmetry, but capable of contributing significantly to the total charge-transfer interaction. Furthermore, differences in relative orientation of donor and acceptor not only alter the relative contributions of these interactions but also change the magnitude of atom–atom repulsions between adjacent molecules within a stack.

Examination of the relative orientations of donor and acceptor found from crystallographic structure determinations of π–π^* molecular complexes[28–31] further emphasises the inadequacy of predictions based on applications of the Overlap and Orientation principle to charge-transfer interactions involving only the highest occupied molecular orbital of the donor and lowest unoccupied orbital of the acceptor (the so-called 'frontier orbitals'). Although the calculations of Prout and Mayoh[32] referred to in chapter 2 reveal that the observed orientations in many of these complexes permit total charge-transfer interaction energies of greater than 90% of the maximum possible in the ideal orientation which maximises the total interaction, there are several examples where HOMO–LUMO overlap contributes nothing to the total interaction. For example, figure 3.17 shows the observed orientation in the anthracene/TCNQ complex, together with

3.17.

HOMO

LUMO

Observed orientation

the orbital symmetries of the anthracene HOMO and TCNQ LUMO. Although the total charge-transfer energy from interaction of all donor and acceptor orbitals (calculated from equations 2.17 and 2.20) is maximised in the observed orientation, it is clear that the overlap of the anthracene HOMO with the TCNQ LUMO is zero in this orientation. Thus, the 'frontier' region of π-donors and π-acceptors must include several donor and acceptor orbitals and not merely the HOMO and LUMO.

In some systems, the observed orientations conflict with the predictions of calculations which include all appropriate orbitals of donor and acceptor. These exceptions fall into two classes, namely systems where strong dipolar orientating influences operate and systems where the orientation required to maximise the charge-transfer interactions grossly conflicts with the requirements for good molecular packing. Typical of the former case are complexes of electron acceptors with bis(8-hydroxy-quinolinato)metal(II) complexes as donors. Figure 3.18 shows the donor–acceptor orientation in the 2:1 complex of 1,2,4,5-tetracyanoben-zene (TCNB) with bis(8-hydroxyquinolinato)copper(II). The strong orien-tating influence of the alignment of a C–CN group above a Cu–O bond, which also places the nitrogen of the CN directly above the copper atom, forces the acceptor molecule to adopt an orientation which provides only 35% of the charge-transfer stabilisation energy provided by the orientation which maximises charge-transfer interactions. The pyrene complexes of TCNB and TCNQ are examples of cases where the optimum orientation for charge-transfer interactions provides grossly unfavourable lattice

3.18. Bis(8-hydroxyquinolinato)copper(II)/2TCNB[33].

packing. Superposition of molecular centres, as required in the ideal orientation, demands stacking of the type shown in figure 3.1(*a*), whereas in fact better packing is obtained with molecules inclined to the stacking axis in these two examples, leading to orientations providing only 61 and 53% of the maximum possible charge-transfer interaction energy. In fact, in many heterosoric structures, the dissimilar shapes and sizes of the donor and acceptor molecules create difficulty in achieving tight molecular packing, and it frequently happens that one of the molecules is not in intimate van der Waals contact with all its neighbours lying in approximately the same plane. In such cases, the molecule may exhibit large in-plane oscillations or orientational disorder, which will be discussed in more detail in chapter 5.

3.4 Charge-transfer salts

In all the electron donor–acceptor systems discussed above, the degree of charge transfer in the ground state is small. However, this is not always true for all donor–acceptor systems, and the combination of a good donor (with low ionisation potential) and a good acceptor (with high electron affinity) can lead to formation of charge-transfer salts with a high degree of charge transfer in the ground state. These salts may be further classified as simple or complex. In simple charge-transfer salts, the donor:acceptor ratio is 1:1, while in complex salts a variety of stoichiometries may occur (for example, DA_2, D_2A, D_2A_3, DA_4, etc.). The degree of charge transfer in the ground state, although large, need not be 100% and can be determined experimentally (for example, using X-ray photo-electron spectroscopy).

The crystal structures of the majority of simple charge-transfer salts between strong π-electron donors and acceptors are composed of heterosoric stacks of alternating donor$^+$ and acceptor$^-$ ions, as would be expected from electrostatic considerations (for example, the tetramethyl-*p*-phenylenediamine salts of TCNQ and *p*-chloranil). However, a smaller number of these simple salts (for example, the tetrathiafulvalene (TTF) and *N*-methylphenazinium salts of TCNQ) form crystals with a homosoric stacking arrangement consisting of stacks of donor cations and segregated stacks of acceptor anions (figure 3.19). These structures are unexpected since they place like charges together in the stacks (although preserving more normal cation–anion interactions between the stacks within the sheets formed by the molecular ions in adjacent stacks). Explanation of their origin is further complicated by the fact that several simple salts have been isolated in two or more polymorphic forms. For example, acetonitrile

solutions of tetraselenafulvalene (TSeF) and diethyltetracyano-
quinodimethane yield either a conducting phase, with segregated
homosoric stacking, or a semiconducting phase, depending on the rate of
cooling during the crystallisation. Such results suggest that the energy
difference between heterosoric and homosoric structures is small, so that
even minor variations in the balance between different intermolecular
forces may suffice to determine which structure has lowest energy.

Since homosoric and heterosoric structures involve totally different
molecular arrangements, so that direct solid-state interconversion of the
two lattices is impossible, the formation of a particular structural type may
well depend on the conditions under which crystal growth is initiated. Once
a particular stacking pattern has been initiated, it will generally be very

3.19. TTF/TCNQ[34].

difficult to prevent subsequent molecules from building onto the initial 'template' in identical fashion. Explanation of the formation of segregated homosoric structures and, more importantly, understanding of possible methods for encouraging their formation by optimisation of crystal growth conditions is thus best approached by considering possible intermolecular interactions which might favour initiation of crystallisation in such structures[35].

There are two main intermolecular interactions which could favour homosoric stack formation: donor–acceptor interaction between D^0 and D^+ or A^- and A^0; and specific attractions between atoms of D^+ and A^- favouring edge-contact of the species rather than face-to-face contact, coupled with the persistence of van der Waals and repulsive forces favouring the homosoric stacks in crystals of the neutral parent D and A molecules.

Donor–acceptor interactions favouring homosoric structures are illustrated by considering the following equilibria in a solution containing TTF and TCNQ:

(i) $\quad TTF + TCNQ \overset{K_1}{\rightleftharpoons} TTF^+ + TCNQ^-$

(ii) $\quad TTF + TTF^+ \overset{K_2}{\rightleftharpoons} (TTF{\cdot}TTF^+)$

(iii) $\quad TCNQ^- + TCNQ \overset{K_3}{\rightleftharpoons} (TCNQ^-{\cdot}TCNQ).$

In dilute solution in acetonitrile, K_1 is such that only a small fraction of the molecules are ionised. TTF^+ will thus more often encounter a neutral donor TTF, forming a $TTF{\cdot}TTF^+$ complex, than one of the few $TCNQ^-$. Similarly, $TCNQ^-$ will more often encounter a neutral acceptor TCNQ and form a $TCNQ^-{\cdot}TCNQ$ complex. These complexes then act as initiation templates for growth of large aggregates with homosoric structure as the solvent evaporates and saturation is reached. In such conditions, TTF·TCNQ crystals elongated along the stacking direction *b* are produced. Neutral donor impurities able to complex with D^+ species may also be important in initiation of homosoric stacking by the charge-transfer mechanism. This has been dramatically illustrated in the case of *N*-ethylphenazinium TCNQ, which normally forms an insulating heterosoric crystal. Addition of about 15 mol% of phenazine TCNQ (which has a neutral ground state) to the solutions prior to crystallisation induces formation of homosoric stacks, via phenazine-*N*-ethylphenazinium complex formation, yielding single crystals of a phase with conductivity 10^9 larger than the normal phase.

Such interactions cannot be invoked to explain the growth of homosoric structures from the vapour phase, for example for TTF·TCNQ. Here specific N ... S interactions appear to favour edge-to-edge contact of TTF and TCNQ, and the second set of factors mentioned above are responsible for homosoric structure. The dominance of edge-on interactions also explains why crystals of TTF·TCNQ grown from the vapour phase are elongated along the edge-to-edge a-direction. More detailed discussion of this problem, and further examples, can be found by the interested reader in reference 29.

In complex charge-transfer salts, homosoric structure formation is easier to understand as these salts formally contain both neutral and ionised donor or acceptor species (for example, quinolinium$^+$(TCNQ)$_2^-$; (Cs$^+$)$_2$(TCNQ)$_3^{2-}$). However, such salts also frequently exist in several structural modifications with very different electrical conductivity properties[36]. In stable ordered structures, periodic electrostatic coupling between anion and cation chains leads to the occurrence of chains of sub-units (dimers, trimers, tetramers, etc.) of cations or anions, in which the overlap between adjacent sub-units is different from that within individual sub-units. This coupling may be destroyed if disorder occurs in one of the chains, and such disordered structures are often associated with high electrical conductivity in the more uniform chains of the non-disordered counter-ions. Failure to recognise the occurrence of different structural modifications in such materials can easily result in wide discrepancies in values of physical properties, such as electrical conductivity, measured by different research groups on what is mistakenly thought to be the same material.

Theoretical interpretation of the observed overlap of molecules within stacks in homosoric charge-transfer salts is difficult[37]. The electrostatic contribution to the lattice energy depends on the degree of charge transfer and the distribution of charge within a stack and within individual molecules, as well as on the precise molecular orientations, and is difficult to evaluate with the accuracy necessary for prediction of minimum-energy structures. As with all molecular crystals, the observed structure is determined by the combined effects of several different intermolecular interactions and there may be several possible structures of closely similar energy. The actual structure adopted may depend on crystallisation conditions, sample purity and temperature, as shown by examples earlier in this chapter. Reliable theoretical predictions of crystal structures of charge-transfer salts are thus impracticable at present, and accurate crystallographic structure determinations are a very important source of primary experimental data, not only for assessing the validity of differing theoretical models for intermolecular interactions but also for aiding interpretation of many other physical properties of molecular crystals.

3.5   Self-assembly

This chapter has described increasingly strong ordering effects reflected in structures of molecular crystals and leading to features such as one-dimensional stacks which have important consequences for the other physical properties of the materials. The ability of organic molecules to assemble themselves spontaneously into complex structures is not limited to the formation of molecular crystals but extends widely into biology. For example, the spontaneous development of the human brain represents the ultimate challenge to those who seek to exploit organic materials in electronics. The ideas presented in the first three chapters of this book concerning crystal growth, intermolecular forces and crystal structures can be extended in several simple ways to guide the development of principles of self-assembly.

Langmuir–Blodgett films

Section 1.2 indicated the importance of slow growth of organic crystals. If molecules could be organised at leisure into well-structured layers and then deposited to form one layer of a film or bulk structure, the problem of separating the stages of depositing and ordering molecules could be solved. Langmuir and Blodgett first explored this approach[38,39] for molecules having hydrophilic and hydrophobic end groups (e.g. long-chain fatty acids). A drop of a solution of the material in a volatile organic solvent that is immiscible with water is deposited onto the surface of very pure clean water in a trough. The solvent is allowed to evaporate and a floating plastic barrier is used to compress the resulting film while the surface pressure is measured. When the molecules come into close contact the surface pressure rises, and the molecules can be forced into close-packed rafts and finally a close-packed layer. If a substrate is then carefully dipped into the trough, the organised molecular layer can be transferred to it in one of the three ways shown in figure 3.20, depending on the nature of the substrate surface and the direction of initial dipping. Several other strategies have also been adopted to control the type of deposition. Thus, the substrate carrier can be made to pass through a seal in the carrier between two troughs covered with different films, while the substrate remains below the surface, so that one material deposits on the downstroke while the other deposits only on the upstroke. In this way, layers of non-linear optic materials can be aligned

with their dipoles all in the same direction, using an optically inactive
alternating spacer layer material. Comparison of the decrease in surface
area per dipping stroke with the geometrical area of the substrate used
provides a measure of the efficiency of transfer of the layer (termed the
'transfer ratio'). This technique has been used successfully not only with
fatty acids but also, for example, with substituted phthalocyanines, a
variety of non-linear optic material (see chapter 10, p. 214) and potential
candidates for molecular rectifiers.

The limitations of the technique have been discussed in some detail in
reference 40. Many of these limitations have their origin in the need for the
deposited material to be soluble in organic solvents for the initial
film-spreading stage. Materials soluble in organic solvents must have quite
low lattice energies. They are therefore likely to form films of limited

3.20. The formation of Langmuir–Blodgett films.

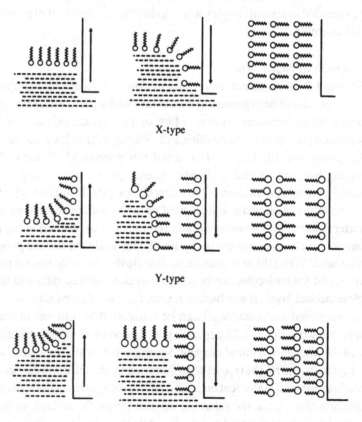

X-type

Y-type

Z-type

mechanical strength, even if well-ordered, and display poor thermal stability. Also they are not very stable to physical disturbances such as those which occur during transfer to the substrate phase, so very slow transfer rates may be required for the highest quality films. The other major limitation is the need for extremely high purity in the materials and in the water used in the trough. Thus, for a film 2 nm thick floating on a water layer 5 cm thick, the ratio of water volume:film volume is 2.5×10^7, and surface-active impurities in the water must be removed to concentrations of the order of 1 part in 10^9 or less if they are not to produce gross effects on the surface film. Impurities are known to increase the chances of defects occurring both in the floating and deposited films, and frequently clean rooms are used to house the entire apparatus to minimise such effects. It is surprising that relatively little attention is paid in the many published papers in this field to the extreme purification of the materials to be deposited by this method compared with the importance which has been widely recognised for this aspect of the study of molecular crystals. Solvent or water incorporation into the deposited films is also an obvious additional problem. These limitations may be of relatively minor significance for room-temperature optical applications and there is a growing realisation of the most appropriate roles for materials deposited in this way.

Use of functional groups to introduce self-assembly properties
The major obstacle to efficient self-assembly in the growth of molecular crystals is the weak intermolecular forces, which do not always align molecules in their equilibrium positions before they are trapped in the growing solid phase. Clearly, therefore, improved self-assembly can be obtained if stronger interactions between the molecules can be achieved to increase the orientating forces *or* if molecules can be designed to move easily until they reach the optimum orientation. Both of these features can be introduced by appropriate substitution. Phthalocyanines provide a good example in both cases. If the periphery of the molecule is substituted with eight long-chain alkyl or alkoxy groups, packing in the crystal is grossly disrupted and frequently such materials do not form good crystals or form structures in which the alkyl chains place the conjugated π-systems of neighbouring molecules as far as 8 Å apart[41]. However, as the temperature is raised, side-chain motion increases and a discotic liquid crystalline state is achieved in which the π-conjugated cores align in one-dimensional stacks (figure 3.21). Electrical conductivity along such stacks has been shown to be high[42]. Alternatively, the long chains may be replaced by crown-ether rings which can complex simple cations. If the

cation fits inside the cavity at the centre of the crown-ether ring, its counter-anion will be attracted to the space between one such complexed cation and another in the next molecule, thus creating linear ionic structures of high stability. If the cation is too large to fit into the cavity it forms a bridge between two crown-ether rings on neighbouring molecules, and the strong ion-dipole forces involved also provide good self-assembly[43]. Other strong dipolar interactions, such as hydrogen bonding or hydrophobic effects (in which molecules assemble for example into micelles, driven by the effect of minimising unfavourable hydrophobic interactions which disrupt the hydrogen-bonded structure of water), are further examples of the use of molecular design to promote self-assembly.

3.6   Analysis of structural trends using crystallographic databases

The great advances in both experimental data collection and data analysis in crystallography since the 1960s have led to a vast number of published structures of molecular crystals – currently approximately 100 000. Crystallographic information is now available from at least five different computerised databases[44]: the Cambridge Crystallographic Database (for organic compounds)[45], the Inorganic Crystal Structure Database[46], the Protein Data Bank[47], the NRCC Metals Crystallographic

3.21. Tetra(18-crown-6)phthalocyanine.

Data File[48] and the Powder Diffraction File[49]. In the case of molecular crystals, it is possible to search the Cambridge Crystallographic Database for particular structural features (defined, for example, by a particular set of atoms connected in a particular way) and also to search for the contact distances and angles between the atoms of such a fragment and atoms of neighbouring molecules. Frequently such a search will reveal a large number of different examples. For example, a search for structures of chloroform solvates involving non-bonded contacts between the C-H group of chloroform and an oxygen atom revealed 100 examples[50]. If the distances and directions of the nearest contact atoms are plotted on a scatter plot, it may be possible to discern patterns which suggest specific interactions that are important in determining the observed crystal packing. (In practice, this can be made easier if each point is replaced by a Gaussian function and the resulting set of overlapping peaks is contoured[51]. Peaks in such a contour map are clearer than the original scatter plots since the eye tends to give undue weight to a few 'outliers'.)

An example of the value of this approach is the study of halogen–halogen interactions in crystal structures. In section 3.1 it was stated that anisotropic repulsion parameters were necessary to model the crystal structure of chlorine. However, it has also been suggested that observed short halogen contacts may be the result of specific attractive forces in certain directions. If the anisotropic repulsion model is used, the number of halogen–halogen contacts relative to contacts with other atoms should be determined by the ratio of the surface area of the halogen atom to that of the whole molecule. However, if specific attractions occur they should favour halogen–halogen contacts and there should be evidence from a survey of structures of halogenated hydrocarbon crystals that the number of such contacts is significantly larger than expected. A study[52] of a total of 229 such structures revealed a significantly higher than expected number of these contacts shorter than the sum of the van der Waals radii, supporting the conclusion from gas-phase molecular beam experiments[53] that specific halogen–halogen attractive forces are involved.

Further examples include studies of the significance of the carbon–hydrogen ratio in aromatic hydrocarbons in determining which of the stacking arrangements discussed in section 3.1 are adopted[54], analysis of the crystal-structure-determining interactions of sulphur atoms[55] and analysis of the structural significance of C–H...O interactions in crystals[56]. The last, particularly, illustrates the importance of searching for specific angles of interaction rather than simply short contacts. The C–N...O interaction is electrostatic and thus falls off much more slowly

48   *Crystal structures*

with distance than van der Waals forces. Thus searching for contact angles corresponding to C–H bonds pointing towards lone-pair directions of the oxygen is probably more useful than simply searching for contacts shorter than van der Waals contact distances.

References

1 A.I. Kitaigorodskii, *Molecular Crystals and Molecules*, New York: Academic Press, 1973.
2 D.E. Williams, *Acta Cryst.*, 1972, **A28**, 629.
3 R.M. Williams and S.C. Wallwork, *Acta Cryst.*, 1968, **B24**, 168.
4 L.O. Brockway and J.M. Robertson, *J. Chem. Soc.*, 1939, 1324; T. Watanabe, Y. Saito and H. Chihara, *Sci. Papers Osaka Univ.*, 1949, **1**, 9.
5 S.C. Abrahams, J.M. Robertson and J.G. White, *Acta Cryst.*, 1949, **2**, 233. D.W.J. Cruickshank, *Acta Cryst.*, 1957, **10**, 504.
6 A.W. Hanson, *Acta Cryst.*, 1963, **16**, 1147.
7 R. Allman, *Z. Krist.*, 1970, **132**, 129.
8 A. Camerman and J. Trotter, *Proc. Roy. Soc. A*, 1964, **279**, 129. J. Tanaka, *Bull. Chem. Soc. Japan*, 1963, **36**, 1237; H.A. Kerr, *Acta Cryst.*, 1966, **21**, A119.
9 G.M. Parkinson, M.J. Goringe, S. Ramdas, J.O. Williams and J.M. Thomas, *J. Chem. Soc. Chem. Comm.*, 1978, 134.
10 J.N. Sherwood, *The Plastically Crystalline State*, New York: Wiley, 1979.
11 W. Nowacki, *Helv. Chim. Acta*, 1945, **28**, 1233.
12 E.D. Stevens, *Mol. Phys.*, 1979, **37**, 27.
13 S.C. Nyburg, *J. Chem. Phys.*, 1964, **40**, 2493.
14 L-H. Hsu and D.E. Williams, *Inorganic Chem.*, 1979, **18**, 79, 2200.
15 S.C. Nyburg and W. Wong-Ng, *Proc. Roy. Soc. A*, 1979, **367**, 29.
16 S.L. Price and A.J. Stone, *Mol. Phys.*, 1982, **47**, 1457.
17 F. Bertinotti, G. Giacomello and A.M. Liquori, *Acta Cryst.*, 1956, **9**, 510.
18 P.J. Wheatley, *Acta Cryst.*, 1955, **8**, 224.
19 P.J. Wheatley, *Acta Cryst.*, 1957, **10**, 182.
20 P.J. Wheatley, *Acta Cryst.*, 1960, **13**, 80.
21 D. Mootz and H-G. Wussow, *J. Chem. Phys.*, 1981, **75**, 1517.
22 S.L. Price and A.J. Stone, *Mol. Phys.*, 1984, **51**, 569.
23 E.O. Schlemper and D. Britton, *Acta Cryst.*, 1965, **18**, 419.
24 D.E. Palin and H.M. Powell, *J. Chem. Soc.*, 1947, 208.
25 J. Gaultier and C. Hauw, *Acta Cryst.*, 1965, **18**, 179.
26 R.G. Little, D. Pautler and P. Coppens, *Acta Cryst.*, 1971, **B27**, 1493.
27 R.E. Long, R.A. Sparks and K.N. Trueblood, *Acta Cryst.*, 1965, **18**, 932.
28 C.K. Prout and J.D. Wright, *Angewandte Chemie (Int. Edn.)*, 1968, **7**, 659.
29 T. Bjorvatten, *Acta Chim. Scand.*, 1966, **20**, 1863.
30 F.H. Herbstein, *Perspectives in Structural Chemistry*, 1971, **4**, 166.
31 C.K. Prout and B. Kamenar, in *Molecular Complexes*, Vol. 1, ed. R. Foster, London: Elek, 1973, p. 151.
32 C.K. Prout and B. Mayoh, *J. Chem. Soc. Faraday II*, 1972, **68**, 1072.
33 P. Murray-Rust and J.D. Wright, *J. Chem. Soc. A*, 1968, 247.
34 T.E. Phillips, T.J. Kistenmacher, J.P. Ferraris and D.O. Cowan, *Chem. Comm.*, 1973, 471.
35 D.J. Sandman, *Mol. Cryst. Liq. Cryst.*, 1979, **50**, 235.
36 G.J. Ashwell, D.D. Eley, M.R. Willis and J. Woodward, *Phys. Stat. Sol. b*, 1977, **79**, 629.

37 B.D. Silverman, *Topics in Current Physics*, 1981, **26**, 108.
38 I. Langmuir, *Trans. Faraday Soc.*, 1920, **15**, 62.
39 K. Blodgett, *J. Am. Chem. Soc.*, 1934, **56**, 495.
40 I.R. Peterson, in *Molecular Electronics*, ed. G.J. Ashwell, New York: Wiley, 1992, p.117–206.
41 I. Chambrier, M.J. Cook, M. Helliwell and A.K. Powell, *J. Chem. Soc. Chem. Comm.*, 1992, 444.
42 J.H. Sluyters, A. Baars, J.F. van der Pol and W. Drenth, *Electroanal. Chem.*, 1989, **271**, 41.
43 O.E. Sielcken, M.M. van Tilborg, M.F. Roks, R. Hendriks, W. Drenth and R.J.M. Nolte, *J. Am. Chem. Soc.*, 1987, **109**, 4261.
44 *Crystallographic Databases: Information Content, Software applications; Scientific Applications.* Bonn: International Union of Crystallography, 1987.
45 F.H. Allen, O. Kennard and R. Taylor, *Acc. Chem. Res.*, 1983, **16**, 146.
46 G. Bergerhoff, R. Hundt, R. Sievers and I.D. Brown, *J. Chem. Inf. Comput. Sci.*, 1983, **23**, 66.
47 F.C. Bernstein, T.F. Koetzle, G.J.B. Williams, E.F. Meyer Jr, M.D. Brice, J.R. Rodgers, O. Kennard, T. Shimanouchi and M. Tasumi, *J. Mol. Biol.*, 1977, **112**, 535.
48 J.R. Rogers and G.H. Wood, in *Crystallographic Databases: Information Content; Software applications; Scientific Applications*, section 2.3. Bonn: International Union of Crystallography, 1987, p.96.
49 J.D. Hanawalt in *Crystallography in North America*, Section D, Ch.2, ed. D. McLachlan Jr. and J.P. Glusker, New York: American Crystallographic Association, 1983, p.215.
50 G.R. Desiraju, *J. Chem. Soc. Chem. Comm.*, 1989, 179.
51 R.E. Rosenfield Jr, S.M. Swenson, E.F. Meyer Jr, H.L. Carrell and P. Murray-Rust, *J. Mol. Graphics*, 1984, **2**, 43.
52 G.R. Desiraju and R. Parthasarathy, *J. Am. Chem. Soc.*, 1989, **111**, 8725.
53 K.C. Jando, W. Klemperer and S.E. Novick, *J. Chem. Phys.*, 1976, **64**, 2698.
54 G.R. Desiraju and A. Gavezzotti, *Acta Cryst.*, 1989, **B45**, 473.
55 G.R. Desiraju and V. Nalini, *J. Mater. Chem.*, 1991, **1**, 201.
56 G.R. Desiraju, *Mol. Cryst. Liq. Cryst.*, 1992, **211**, 63.

4

Impurities and defects

4.1   Effects of impurities on crystal properties

Even the most carefully purified and grown molecular crystals inevitably contain residual impurities and defects. The impurities may be neutral molecules whose properties (shape, size, solubility, etc.) render them difficult to separate from the host material (e.g. anthracene in phenanthrene), or species produced as a consequence of the reactivity of the material (e.g. tetracyanoquinodimethane (TCNQ) readily forms Na^+TCNQ^- in contact with glassware). The effects of such impurities on the properties of the crystals arise as a result of structural or electronic perturbations associated with the impurities.

From the discussion of intermolecular forces and crystal structures in chapters 2 and 3, it follows that an impurity molecule, by virtue of its different shape and electronic structure, will interact with neighbouring molecules of the host lattice in a way which is different from the interactions between a host molecule and other surrounding host molecules. Hence the host molecules surrounding an impurity have energies and positions which differ from those of the normal host lattice to an extent depending on the difference between the shape and electronic structure of the impurity and the host molecules. These local effects surrounding impurity molecules are commonly called *X-traps*, since the altered energies of the molecules frequently result in the trapping (or localisation) of electronic excitations in the crystal (see chapter 6). In other cases, impurity molecules may lead to gross structural changes. For example, as discussed in chapter 3, the addition of phenazine to solutions of *N*-ethylphenazinium/TCNQ alters the structure of the complex crystallising from solution from a heterosoric semiconducting phase to a homosoric conducting phase. Fortunately, such large structural changes do not normally occur as a result of the presence of impurities in the concentrations of less than about 100 ppm commonly achieved by the careful application of the techniques described in chapter 1. For example, 15% of phenazine is required in order to produce the gross effect referred to above for *N*-ethylphenazinium/TCNQ. Nevertheless, a check should always be made for unexpected impurities (e.g. from side-reactions, wet solvents, etc.) before sophisticated hypotheses are developed to explain the formation of multiple phases or phases whose structures and properties are anomalous.

Electronic effects of impurities are threefold:

the formation of exciton or charge-carrier traps;

the creation of charge-carriers by ionisation of impurities;

the disruption of the regular band structure associated with the normal lattice periodicity.

Impurities with orbitals lower in energy than those of the molecules of the host lattice will be energetically-favoured sites for electronic excitation energy or charge carriers and will therefore act as exciton or charge-carrier traps, with consequent effects on optical and electrical properties of the crystals (described in chapters 6 and 8). Conversely, impurities which are more easily ionised than molecules of the host will facilitate charge-carrier generation and enhance semiconductivity. The energy (ΔE) required to create a pair of separated positively and negatively-charged molecular ions in a neutral molecular crystal is given by[1]

$$\Delta E = IP - EA - P^+ - P^- + \Delta W_f \qquad (4.1)$$

where IP = gas-phase ionisation potential

EA = gas-phase electron affinity

P^+ and P^- = polarisation energies of the resulting ions (i.e. their interaction energies with the surrounding lattice of neutral molecules)

and ΔW_f = lattice stabilisation of the molecular ground state.

Thus, impurities with either lower ionisation potential or higher electron affinity than those of host-lattice molecules will lead to lower activation energies for charge-carrier generation, and experimental determination of this activation energy from the temperature-dependence of semiconductivity may in favourable cases reveal whether the observed conduction is intrinsic or extrinsic (see chapter 8). Finally, in certain classes of molecular crystals, particularly charge-transfer salts with homosoric structures, the presence of energy bands resulting from overlap of the molecular orbitals of adjacent molecules is an important feature determining the electronic conduction properties of the solid. The formation of such bands requires a regular periodic lattice structure and good overlap between orbitals of adjacent molecules. Impurities disrupt the regular lattice structure and may have little overlap with orbitals of neighbouring molecules, so they lead to scattering of charge-carriers moving in the band structure. These effects, also discussed further in chapter 8, are particularly significant in solids where the structure is highly anisotropic, leading to one-dimensional band structures (as in homosoric charge-transfer salt structures) since such band structures provide no alternative routes for charge carriers to by-pass the discontinuities resulting from impurities.

4.2   Types of structural defect

These structural imperfections associated with isolated impurity molecules are examples of point defects in molecular crystals. The other major types of structural defect are line defects, planar defects, grain boundaries and structural disorder.

Point defects

Point defects are defects that are localised around a point and not extended in one or more dimensions. Other examples of point defects in molecular crystals include vacancies, interstitials and mis-oriented molecules. Vacancies, in which a molecule is simply missing from a lattice site leaving a void, commonly have a creation energy comparable with the enthalpy of sublimation and are therefore small in number in molecular crystals, in contrast to the situation in, for example, ionic solids[2]. Similarly, efficient lattice-packing and the steep repulsive potential between molecules at short distances of approach make the formation of interstitials (i.e. molecules which lie in between two or more normal adjacent lattice sites) thermodynamically unfavourable. Defects arising from mis-oriented molecules are more common since mis-orientation frequently moves only a small part of the molecule from its normal position. For example, 9-cyanoanthracene has the crystal structure[3] shown in figure 4.1, with molecules stacked along the c-axis. Within a given stack, all the cyano groups normally point in the same direction, but an in-plane rotation of $180°$ leaves the position of the anthracene atoms unchanged, with only the cyano group moved from one side to the other. Molecules may accidentally arrive in mis-oriented positions during the process of crystal growth, or they may undergo re-orientation by rotating on their lattice sites. The latter process, discussed further in chapter 5, has an associated activation energy depending on the interactions between a molecule and its neighbours during the rotation. Thus, mis-oriented molecules arising from errors in crystal growth should contribute a temperature-independent number of point defects, while those arising from on-site rotation will occur more commonly at higher temperatures.

Line defects or dislocations

Line defects or dislocations are of two idealised types: edge dislocations and screw dislocations[4]. In edge dislocations an extra plane of molecules is inserted part-way into the lattice, as shown in figure 4.2. The line along the leading edge of the inserted plane is known as the dislocation line and the plane containing this line and the normal to the inserted plane is the slip plane. The region of strain near a dislocation is referred to as the core of the dislocation. The magnitude of this strain is concisely expressed in terms of the Burgers vector b, as shown in figure 4.2. Any circuit such as $ABCDA$, which would be a closed loop in a perfect crystal, will become $A'B'C'D'E'$ and fail to close in a region of the crystal containing a dislocation. The lattice vector (in this case directed $E'A'$) required to close the loop is known as the Burgers vector b. For edge dislocations the Burgers vector is

4.1.

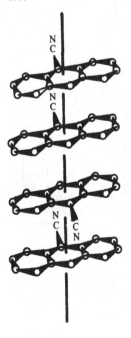

perpendicular to the dislocation line. In a screw dislocation, one part of the crystal slips by one lattice spacing along the slip-plane, relative to the part of the crystal below the slip-plane, as shown in figure 4.3(*a*). The resulting arrangement of lattice sites around the dislocation line is shown in figure 4.3(*b*) and is a spiral; hence the name screw dislocation. In this case the Burgers vector is parallel to the dislocation line. Mixed dislocations with Burgers vectors having both edge and screw dislocation components are also possible, and dislocation lines are therefore frequently curved.

4.2. Edge dislocation.

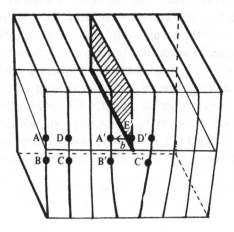

4.3. Screw dislocation.

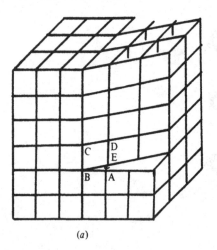

(*a*)

(*b*)

Dislocations may also be classified according to the magnitude of their Burgers vectors. Thus a dislocation is said to be perfect (or of unit strength) when the Burgers vector has the same magnitude as a unit lattice vector. Multiple-strength dislocations have Burgers vectors several times the magnitude of the unit lattice vector, while partial dislocations have Burgers vectors less than the unit lattice vector. The general notation for a dislocation is thus in three parts:

(slip-plane) fractional lattice displacement equivalent to Burgers vector, if partial [direction of slip]; for example: $(010)\frac{1}{2}[100]$.

It has been shown that the total strain energy associated with a dislocation is proportional to the square of its Burgers vector. Therefore, it is energetically favourable for dislocations to dissociate into a series of partial dislocations with smaller Burgers vectors. The total strain energy of a dislocation is the sum of the core energy and the extended region of elastic strain in the lattice surrounding the core. It has been estimated that a region of radius 10^{-4} cm around the core of a dislocation accounts for the core energy plus approximately half of the associated elastic strain energy of the surrounding lattice, although clearly this figure is very approximate, varying according to the material and the direction of the dislocation.

The dislocation density is defined as the number of dislocation lines passing through unit area of a solid. In vapour-grown molecular crystals, dislocation densities may be as low as 10 cm^{-2}, while in melt-grown crystals they may be as high as 10^5–10^7 cm^{-2}, with solution-grown crystals lying between these extremes. The number of anomalous lattice sites associated with dislocations is larger than the dislocation density, as a result of the extended nature of the surrounding lattice distortions. Thus a dislocation density of 10^6 cm^{-2} typically yields an anomalous site density of 10^{14} cm^{-3}. Nevertheless, the entropy change associated with the formation of a dislocation is small compared with the strain energy involved, so that dislocations are thermodynamically unstable and ideally should be removed by sufficient annealing. One example of a mechanism for annealing out dislocations is shown in figure 4.4, in which two edge dislocations on opposite sides of a slip-plane migrate laterally until the inserted planes of molecules align.

Dislocations, by virtue of the associated regions of strain and anomalous sites in which molecules lie off their equilibrium-perfect lattice sites, influence molecular motion, the migration of charge and electronic excitation energy and the chemical reactivity of molecular crystals, as will be discussed in later chapters. They may also influence the distribution and concentration of impurity molecules in the crystal. For example, consider the edge dislocation in figure 4.2. The lattice is compressed above the slip-plane by the insertion of the extra lattice plane, and this region will thus preferentially incorporate any impurity molecules which occupy a smaller volume than the host molecule. Conversely, the region below the slip-plane has space available for impurities of larger molecular volume than the host. Thus, impurities frequently congregate at dislocations in preference to more perfect regions of the lattice. This may, for example, facilitate the incorporation of solvent molecules into solution-grown crystals which, as mentioned above, have higher dislocation densities than vapour-grown crystals.

Planar defects or stacking faults

These defects occur when the regular lattice packing is disturbed in two dimensions, rather than along a single line as in the case of dislocations. The simplest example occurs for close-packed lattices of effectively spherical molecules, such as the plastic crystalline phases referred to in chapter 3. A close-packed layer of spheres as shown in figure 4.5 has two sets of positions

4.4. Migration of edge dislocations during annealing.

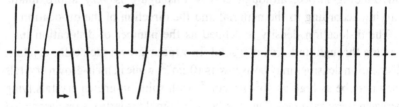

4.5.

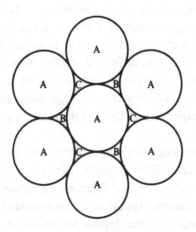

(*B* and *C*) in which the spheres of the next layer may fit. In the face-centred-cubic phase normally adopted by plastic crystals, the layer sequence is ... *ABC ABC ABC* ..., while in the hexagonal close-packed structure the sequence is ... *ABABAB* .... Within the face-centred-cubic structure, occasional sequence faults such as ... *ABC ABC BC ABC ABC ABC* ... can occur with very little energy penalty; these are planar defects.

In practice, stacking faults are observed by electron microscopy (as described later in this chapter) to occur in localised regions, for example as ribbons threading through the solid, rather than as planes extending throughout the bulk of the solid. This observation can be explained by considering the process of a perfect dislocation dissociating into two partial dislocations. Such a process, in which the sum of the Burgers vectors of the two partial dislocations is the same as the Burgers vector of the original perfect dislocation, is energetically favourable as explained above, and the region separated by the two partial dislocation lines corresponds to a stacking fault ribbon. Figure 4.6 shows how dissociation of a perfect dislocation of a Burgers vector along [100] in a 9-cyanoanthracene crystal into two partial dislocations of Burgers vectors one quarter of the unit translations along [21ω] and [2$\bar{1}\bar{\omega}$] leads to the molecules in broken lines

4.6. Planar defect structure in 9-cyanoanthracene.

in the layer above the plane of the paper. Half of these molecules, interestingly, lie centrically related to their neighbours in the next sheet, and this is thought to account for the unexpected observation that solid-state photodimerisation of 9-cyanoanthracene yields the *trans* product and not the expected *cis* product[5]. (The energy difference of the abnormally stacked region serves preferentially to trap the electronic excitation energy arising from illumination, at sites with orientation favouring the *trans* product.) Partial dissociation of two other types of dislocation also produces similar orientations in these crystals. Although mis-orientation of individual 9-cyanoanthracene molecules, as depicted earlier in this chapter, could produce similar orientations leading to some *trans* dimer formation, the number of such isolated point defects is insufficient to account for the observed yield of the *trans* product.

Crystal twinning and grain boundaries

The formation of twinned crystals may also arise as a result of stacking faults, as shown earlier (figure 3.5, p. 27). In general, twins in which the two parts meet along a definite plane, called the *composition plane*, are known as *contact twins* (A/B), and their orientations may be related by reflection across a common lattice plane known as the *twin plane*. In *polysynthetic twins*, the twinning is repeated ($A/B/A/B/A$ etc.). For both of these types of twinning, two diffraction patterns corresponding to regions with orientations A and B will be observed superimposed in X-ray diffraction investigations. *Multiple twins* are composed of more than two parts ($A/B/C$ etc.) and thus produce more than two superimposed diffraction patterns. In cases where stacking faults of the type required to produce twinning are energetically unfavourable, for example on the grounds of lattice-packing considerations, twinning may still occur across a *low-angle grain boundary* between the two parts of the twin. Such a grain boundary is shown in figure 4.7, where two parts of the crystal grow at an angle θ to each other by formation of a series of edge dislocations of Burgers vector b a distance D apart along the boundary plane. Similarly, *twist low-angle grain boundaries* may be formed from a series of screw dislocations.

It is thought that most real crystals are in fact composed of mosaic blocks each a few hundred to a few thousand lattice repeat units across, inclined at a few minutes of arc to each other and interfaced by very low-angle grain boundaries. The evidence for this is from the widths of X-ray diffraction peaks, which theoretically should be only a few seconds of arc for a perfect single crystal but which in practice are a few minutes of arc because of the spread of mosaic block orientations. At the other extreme, in highly polycrystalline material, such as compacted powder pellets of molecular crystalline materials, high-angle grain boundaries occur in which $\theta > 30°$ and the dislocation model breaks down as the separation required becomes comparable to the lattice spacing[6]. Little is known of the structure of such boundaries on a molecular scale (e.g. whether they are sharp discontinuities or more gradual structural changes with almost-amorphous regions linking the two differently oriented crystallites). However, their presence is clearly associated with significantly higher defect and dislocation contents than in single crystal samples. Furthermore, random orientation of individual crystallites in such compacted pellets means that the ability to measure the influence of crystal structure anisotropy on physical properties is lost. The use of well-annealed single-crystal samples in physical measurements on molecular crystalline materials is therefore strongly preferred. Even in these samples, however, the weak intermolecular forces can permit structural disorder, which is therefore an inherent defect in significant numbers of molecular crystals.

Structural disorder
Structural disorder can occur whenever the crystal lattice packing permits a molecule to adopt two different orientations on its lattice site with similar

4.7. Grain boundary.

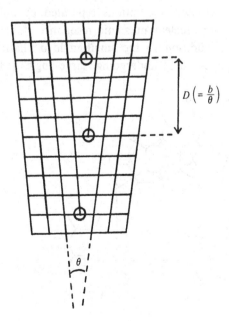

energies. This can occur in homomolecular crystals if the molecular shape is not too far from symmetrical. For example, azulene[7] and 1,4-bromo-chlorobenzene[8] both crystallise in space groups which require that the molecules occupy centrosymmetric sites. Since the molecules themselves are not centrosymmetric, this means that each site is apparently randomly occupied by molecules in two orientations which are related by a centre of symmetry. X-ray diffraction studies determine only the average structure (unless the disordered molecules are arranged in some ordered way in one or two dimensions, or in regions each of which exhibits some short-range order, in which case the order is reflected in characteristic diffuse-scattering structure). Hence from X-ray studies it is not possible to determine if the disorder is static or dynamic, although other techniques such as nuclear magnetic resonance relaxation studies do permit this distinction to be made, as will be discussed in chapter 5. Disorder is generally more common in crystals composed of two or more distinct molecular species, where the lattice packing may be more strongly influenced by the requirements of one of the molecules than by those of the other. For example, in many weak $\pi-\pi^*$ donor–acceptor complex crystals, the acceptor lattice is ordered whereas the donor is disordered (e.g. phenanthrene/1,2,4,5-tetra-cyanobenzene (TCNB)[9], figure 4.8). In some cases, the disorder may be difficult to distinguish from thermal motion if the two disorder orientations only involve a small-angle in-plane rotation (e.g. anthracene/TCNB at room temperature[10], where the anthracene appears to be undergoing in-plane oscillation between two orientations separated by only 8.6° in-plane rotation). However, nuclear magnetic resonance studies have shown that frequently in addition to the small-angle oscillation ($\theta°$) the disordered molecule is also undergoing large-angle $(180-\theta°)$

4.8. Phenanthrene/TCNB.

oscillation[11,12]. Atom–atom potential calculations have been used to calculate the energy barriers to such molecular rotations to compare with the experimental values. Cases are known where such rotation does (e.g. naphthalene/TCNB) and does not (e.g. phenanthrene/TCNB) occur. In the case of homosoric charge-transfer salts, the occurrence of disorder in the cation or anion sites can have very important consequences for the electrical conductivity properties (chapter 8), and there are many well-documented examples. For example, careful examination[13] of X-ray diffraction photographs of *N*-methylphenazinium-TCNQ revealed faint but systematic streaks in positions which indicated the structure shown in figure 4.9. The arrangement of *N*-methylphenazinium and TCNQ is regular in each *a–b* layer, with the lattice periodicity along the *a*-direction twice that originally reported, but each such layer has no correlation with adjacent layers.

4.3   Characterisation of impurities

Full characterisation of a molecular crystal by determination of the nature, concentration and distribution of all the impurities and defects which it contains is a major undertaking that is very rarely carried out. However, suitable methods do exist for the various stages of such characterisation for most materials, and the problems of their application are generally related to the time required rather than to the inherent feasibility of the project. Since very small impurity concentrations (in the ppm range) can have appreciable effects on the physical properties of

4.9. *N*-Methylphenazinium-TCNQ.

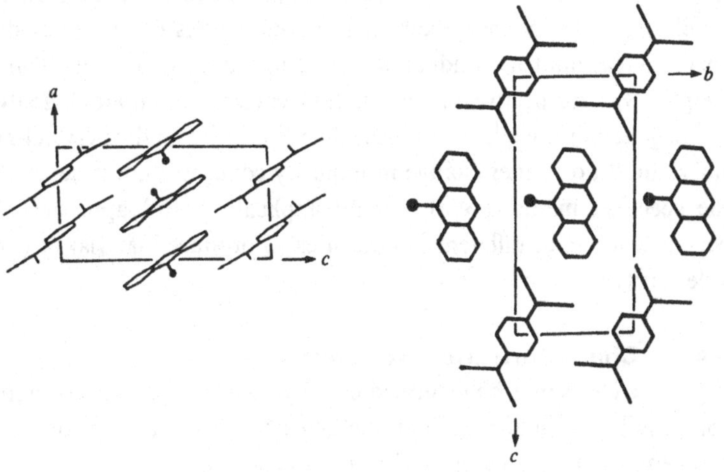

crystals, conventional chemical analytical methods are of limited value as criteria of purity.

In some cases, visual examination of the product of a purification process can identify the purest material. Thus, zone-refining of aromatic hydrocarbons often produces tubes containing brown zones of impurities at one or both ends, with pure, often colourless, material in the centre. Similarly, entrainer sublimation produces a band of the purified material which is often of a different colour or crystal form from the less or more volatile impurities in other regions of the apparatus. Examination under ultra-violet illumination may reveal further impurities which, though not coloured, do fluoresce. Clearly these observations can be made quantitative by appropriate absorption or fluorescence spectroscopic measurements (see chapter 6). Vapour-phase chromatography is also an excellent technique for detecting small amounts of impurities. Comparison of retention times with those of known materials can be a guide to the nature of the impurities using this method, which may also be used in tandem with other analytical techniques such as mass spectrometry to provide a very versatile and powerful system for identification and analysis of wide ranges of impurities. Minor adaptations can sometimes greatly increase the effectiveness of this technique for molecular crystal materials. Thus the use of liquid-crystal coatings (e.g. *N,N'*-bis(*p*-methoxy-benzylidene)α,α'-bi-*p*-toluidine, BMBT)[14] on standard column-packing materials facilitates the separation of mixtures of aromatic hydrocarbons. In other cases, the presence of impurities can be inferred from detecting their influence on other physical properties. Thus, Na^+TCNQ^- impurities in TCNQ crystals can be inferred from electron-spin resonance (ESR) observations and from extrinsic regions in the temperature-dependence of semiconductivity. Similarly, impurity charge-carrier trapping centres can be studied using space-charge-limited conduction or drift-mobility measurements (see chapter 8). Irrespective of the method chosen for impurity analysis, a useful general principle is to repeat the purification and/or crystal-growth stages until no further change in impurity concentration is detectable by the specified method. With careful application, this approach should permit workers in different laboratories to produce material of comparable purity.

4.4   Characterisation of defects

Defects may be identified directly by optical and electron microscopy and by diffraction methods including X-ray topography, or indirectly by chemical etching or by positron annihilation.

Optical methods

Individual crystals with well-developed flat faces can be characterised by examination under an optical microscope. Measurement of interfacial angles using an optical goniometer and subsequent plotting of stereographic projections may permit identification of the crystallographic planes corresponding to crystal faces. If the crystal is rotated on a microscope stage between crossed polarisers, any departure from uniform extinction will reflect structural discontinuities (twinning, grain boundaries, etc.). For details of the other information (e.g. location of optical axes) available from optical microscopy of crystals (e.g. using convergent polarised light) but not directly relevant to identification of defects, the reader is referred to a standard text on optical crystallography[15].

Optical characterisation of defects may be extended in scope by the cheap, rapid and widely-applicable technique of chemical or physical etching. The principle is simple: molecules in anomalous sites surrounding defects dissolve or evaporate more readily than molecules in normal lattice sites. Thus etching or evaporation leads to pits in the surface at defect sites, and these pits may be located and counted under a microscope (figure 4.10).

4.10. Dislocation etch-pits (acetone 5 s etch at 273 K) developed around a micro-hardness indentation mark on a (110) surface of a pentaerithritol tetranitrate crystal. The pits are formed at the emergent ends of dislocation loops punched into the crystal by the indentation. The loops lie in the (110) slip plane. (Courtesy Prof. J.N. Sherwood; c.f. P.J. Halfpenny, K.J. Roberts and J.N. Sherwood, *J. Materials Science*, 1984, **19**, 162.)

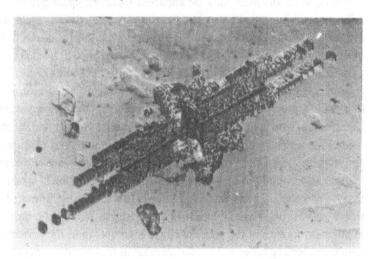

The principal disadvantage of this technique is that it is destructive, but this can be overcome by the use of a cleaved single crystal. Thus, for example, the influence of dislocations on reaction rates for photodimerisation in anthracene has been studied by reacting one cleaved face and etching the other face which was formerly in direct contact with it[16]. A good correspondence was observed between reaction-product distribution and etch-pit distribution, suggesting that dislocations enhance either the reaction rate or the crystallisation of the reaction products. The possibility that the operation of cleaving might induce more defects in one or the other of the two cleaved sections can often be eliminated by observing a good correspondence between the distribution of etch pits on the two faces exposed by cleavage. Typical etching agents that have been used for molecular crystals include concentrated sulphuric acid, methanol, ethanol + 1% glycol, acetone, dimethylformamide, cyclohexanone and acetone/water mixtures.

X-ray topography

Optical studies can detect only certain types of defect (twins, grain boundaries) when they occur on a scale large enough to be visible and in crystals which do not absorb light strongly. A method which avoids these limitations is X-ray topography, one of the most powerful and generally applicable techniques for assessment of crystal quality[17]. Normal single-crystal X-ray diffraction studies provide information on some crystal defects (e.g. twinning can be seen as two or more superimposed diffraction patterns, while disorder may be apparent from the space group or from diffuse scattering streaks as mentioned above) but do not yield data on dislocations, stacking faults, etc. However, the intensity of the diffracted beam from a particular set of crystal planes is perturbed if the incident beam falls on a region containing defects. For example, a boundary between two regions of a crystal which have slightly different orientations (e.g. a low-angle grain boundary) will lead to a region of contrast in the reflected X-ray beam, which arises in three ways depending on whether the incident beam is spectrally continuous or monochromatic, and collimated or convergent, as shown in figure 4.11. This is known as *orientation contrast*. Defects also alter diffracted-beam intensities via *extinction contrast*, which may be revealed as a shadow cast by the defect scattering diffracted beams which pass it, or as an enhancement of intensity around the defect if the angular divergence of the incident beam is greater than the perfect-crystal angular reflecting range so that the increase in this angular reflecting range resulting from the slight mis-orientation around the defect increases the proportion of the incident beam diffracted. (The former effect is known as the dynamical image, while the latter is the direct or kinematic image.) Thus, if a crystal is uniformly exposed to an X-ray beam across its whole surface, examination of the resulting reflection will provide an image of defects in black and white contrast. Figure 4.12 shows two experimental arrangements for such exposures. In the Berg–Barrett technique, the crystal is examined in a reflection arrangement with shallow-incidence X-rays; the image is predominantly a surface image since the X-rays penetrate only a few microns into the crystal surface in this geometry. In the Lang technique, the focussed, collimated X-ray beam is diffracted as it passes through the crystal, and the desired reflection is isolated using a slit. Both the crystal and the film are synchronously traversed across the X-ray beam to obtain

4.11. Origins of contrast in topography.

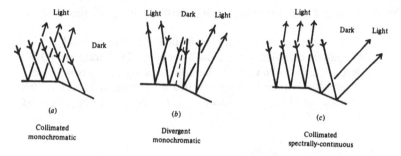

(a)
Collimated
monochromatic

(b)
Divergent
monochromatic

(c)
Collimated
spectrally-continuous

4.12. (*a*) Berg–Barrett and (*b*) Lang cameras.

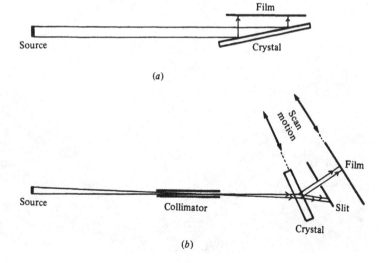

an image of the whole sample, whose thickness can be up to a few millimetres in favourable cases, yielding information on bulk defects.

These techniques require long exposures (up to two days, typically) and very careful alignment of the crystal to within a minute of arc throughout the exposure. Recently, however, the availability of synchrotron radiation facilities has made possible exposure times of less than one minute with crystals whose orientation is far less critical (typically within about one degree of the ideal selected orientation) because of the high-intensity continuous radiation available[18]. These advantages may be offset in some cases by increased radiation damage in organic solids exposed to such high radiation intensities. Figure 4.13 is one example of a topograph.

Electron microscopy

Electron microscopy[19] is another direct method for identifying defects, one which also suffers from the problem of radiation damage[20], this time from

4.13. X-ray topograph of a (110) section of a pentaerithritol tetranitrate crystal showing growth horizons (G), dislocations (D1) developing from the seed (S) and dislocations (D2) and twin bands (T) developing from inclusions (I) (Cu K_α (radiation). (Courtesy Prof. J.N. Sherwood; c.f. P.J. Halfpenny, K.J. Roberts and J.N. Sherwood, *J. Crystal Growth*, 1984, **67**, 320.)

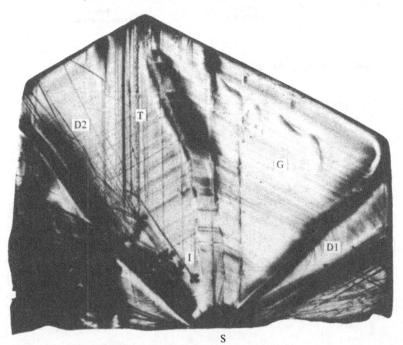

inelastic scattering of electrons leading to energy transfer to the molecular crystal. While increasing the accelerating voltage applied to the electron beam reduces this inelastic scattering, it also reduces elastic scattering and the phosphors used to detect the image also become less efficient for high-energy electrons. Use of low sample temperatures, while not altering the scattering cross-section, does reduce radation damage by restricting migration of radicals formed by the electron beam and by minimising effects resulting from sample volatilisation. Appropriate chemical substitution, for example replacement of hydrogen by chlorine in aromatic compounds, is also found to reduce radiation damage for similar reasons. Finally, the use of thin samples reduces radiation damage, since the electron beam is slowed down as it passes through the material, and the lower layers of a thick sample are therefore exposed to a higher proportion of low-energy more-damaging electrons. For transmission electron microscopy, samples of less than about 5000 Å thick are needed for organic materials, and for materials which do not naturally crystallise as thin plates, chemical thinning of sectioned large crystals using a jet-polisher may be necessary. The lowest possible electron beam intensity should be used, since sudden exposure of organic solids to concentrated electron beams can result in sublimation and dendritic crystal growth. Figure 4.14 shows how the objective lens focusses both transmitted and diffracted beams from the sample. The electron microscope can thus be used to examine the diffraction pattern produced by the small sample areas illuminated by the

4.14.

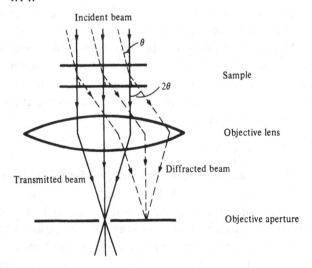

electron beam, as well as to produce highly-magnified direct images of this area. In the latter case, the objective aperture shown in figure 4.14 can be used to select which of the diffracted beams are allowed to contribute to the diffracted image. Contrast arising from defects shows up in the final image from a particular diffracted beam in an analogous way to that in X-ray topography, in high magnification. In particularly suitable cases (e.g. phthalocyanines) molecular images can be obtained by electron micros-copy, which permit direct observation of the molecular positions around defects and phase boundaries, etc[21,22]. The introduction of electron microscopes with superconducting lenses and facilities for maintaining the sample at temperatures around 4 K may well extend the very limited range of materials for which studies at this ultimate level of detail can be carried out[23]. The obvious limitations of electron microscopy are high cost, the time needed and the limitations imposed by sample requirements.

Scanning probe microscopy
The most direct way to study surface structure and defects on an atomic scale would ideally be to use the same approach applied by blind people: feel the surface contours and hence deduce the shape. This is the principle of scanning probe microscopy. Three problems immediately spring to mind: how to control the scan sufficiently finely, how to maintain a constant gap between the tip and the surface, and how to obtain a sufficiently fine tip to probe on an atomic scale. The first of these is solved using piezoelectric ceramic scanners, which combine very fine movement with electrical readout of position. The second can be solved in a variety of ways, the first of which led to the development of scanning tunnelling microscopy (STM), for which Binnig and Rohrer were awarded a Nobel Prize in 1986[24]. A fine metal tip is brought to within 8 Å (typically) of the surface, with a small voltage (a few millivolts to a few volts) applied between the tip and the surface. Electrons tunnel between tip and surface, the tunnelling current depending on the separation (S) between the tip and surface according to the Fowler–Nordheim equation:

$$I_T \propto V_T/S \, \exp(-A\phi^{\frac{1}{2}}S)$$

where A is the Fowler–Nordheim constant $(1.025(\text{eV})^{-\frac{1}{2}}\text{Å}^{-1})$; ϕ is the average of the work functions of the tip and surface and V_T is the applied voltage. This equation predicts that the current will change by a factor of 10 if the probe–sample gap changes by as little as 1 Å. Thus, using a feedback loop to maintain constant tunnelling current will maintain a constant separation, provided the work function of the surface is uniform at all points. This assumption is not always true, and the technique, therefore, gives a map which convolutes changes in surface topography and work function. This map is displayed on a computer monitor in which the plane of the sample is the XY plane of the screen while the third dimension is imaged using a range of colours for different heights, Z. Clearly this technique can only work if the sample is electrically conductive, although it has been successfully used with thin samples of insulating materials on conductive substrates. The problem of how to obtain a suitable probe tip is less serious than might be imagined, and a fine wire cut on a slant with a sharp pair of scissors is often used. The reason for success is that the single atom nearest to the surface will dominate the observed tunnelling current, because of the very rapid fall off in current with distance from the surface referred to above. Usually if the tip is unsuitable the derived image will be clearly unusable, though there have been reports of apparent superlattice-like images arising from contributions from more than one atom of the probe.

An alternative approach is to mount the probe on a springy cantilever beam and maintain a constant force on the spring, as measured by the deflection of a laser beam reflected from the tip of the spring. (The sample is therefore moved rather than the probe tip in this technique.) The spring force constant must be less than that of the equivalent 'spring' between atoms in the solid, to avoid the possibility of the spring actually pushing atoms aside. The required constant, $\sim 1 \, \mathrm{Nm^{-1}}$, is equivalent to that of a piece of cooking foil approximately $4 \times 1 \, \mathrm{mm}$. This technique, known as atomic force microscopy (AFM)[25], does not require that the sample is conductive. A whole range of variants on this approach has also been explored[26]. For example, if the tip on its cantilever mount is brought not quite so close to the surface (2–20 nm), an attractive force (approximately 1000 times smaller than the repulsive forces probed by AFM) is experienced. If the cantilever is mounted on an oscillator driven at just above the assembly's lowest mechanical resonance (typically 50 kHz), an amplified oscillation of the tip occurs which can be detected by laser interferometry. As the tip approaches the surface, the attractive forces increase, giving the effect of softening the spring so that the natural resonance frequency is now lower and hence further from the driving frequency. The oscillation thus moves off resonance and the amplitude is reduced. This is known as laser force microscopy (LFM).

These techniques all require anti-vibration mountings, and all suffer from a common problem, namely that it is impossible to 'zoom-in' from a low magnification view to the final image, so it is difficult to determine exactly which area of the sample is being examined and whether the image is typical of the majority of the surface. This problem is overcome by examining a number of different points on the surface. Features common to all points may be considered representative. Clearly, if a region contains a surface defect this will be immediately obvious in contrast to the 'normal' image, and detailed information about defects and other surface modifications (such as surface changes associated with organic solid/gas reactions[27]) can be obtained on an atomic scale.

Scanning probe microscopy is substantially cheaper than electron microscopy of the same resolution but is complementary to electron microscopy in the sense that it provides surface information on samples of any thickness, whereas transmission electron microscopy requires ultra-thin samples. Other significant differences are that electron microscopy necessarily has the sample in a vacuum whereas this is not necessary for STM etc.; STM gives true three-dimensional information whereas the third dimension is at best impressionistic for electron microscopy; and scanning probe microscopies cause less damage to the sample than that inflicted by the electron beam.

Positron annihilation

This technique depends on the fact that the lifetime of positrons (e^+), produced by radioactive decay of ^{22}Na, depends on the probability of annihilation by collision with electrons of a solid material. This probability can be measured by measuring the positron lifetime by use of a coincidence counter, which measures the frequency of occurrence of the simultaneous emission of the two 0.51 MeV X-ray photons which accompanies the annihilation process. The probability depends on the density of electrons in the vicinity of the positron and is therefore sensitive to point defects as well as to the proportion of free space in the lattice, as reflected by the packing coefficient. For example, figure 4.15 shows the expected inverse correlation between positron lifetime and packing coefficient[28], while figure 4.16 shows the increased probability of decay, as reflected in higher X-ray emission intensity, as the small molecules anthracene and 2,3-benzofluorene are doped into the lattice of the larger host molecule *p*-terphenyl[29].

Figure 4.17 shows the variation of the lifetime of one component of the positron lifetime spectrum for succinonitrile as a function of temperature[30]. At low temperatures the material has a brittle phase in which the positron lifetime increases slowly with temperature, probably as a result of thermal expansion of the lattice, until the material undergoes a transition to a plastic phase. This phase has a lower heat of sublimation and a correspondingly higher vacancy concentration, giving a longer positron lifetime. Analysis of such data has been used to derive experimental vacancy-formation energies for plastic phases, which are found to be typically in the range 0.5–1 times the heat of sublimation.

The limitations of positron annihilation are cost and the difficulty of deconvoluting the data for cases where several lifetimes exist and linking these lifetimes to specific physical processes unambiguously. These

4.15. Re-drawn after M. Eldrup[28].

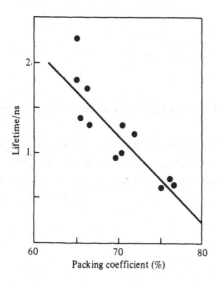

4.16. Intensity of 0.51 MeV X-ray emission as a function of concentration of anthracene or 2,3-benzofluorene in *p*-terphenyl.

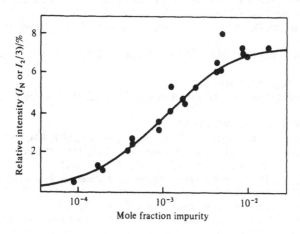

4.17. Temperature-dependence of positron lifetime for succinonitrile.

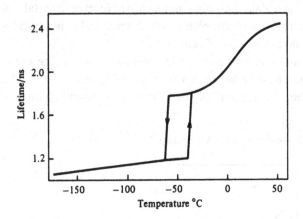

limitations have resulted in a very few reports of studies using this technique in recent years.

There remain a vast number of molecular crystals which are increasingly being considered for electronic and optical applications and whose defect properties may adversely influence their performance in such applications. The scope for wider use of the techniques described in this chapter, therefore, is still large.

References

1 L.E. Lyons, *J. Chem. Soc.*, 1957, 5001.
2 A.V. Chadwick, in *Mass Transport in Solids*, NATO Advanced Study Institute Series B, **97**, Ch. 11, ed. F. Bénière and C.R.A. Catlow, New York: Plenum, 1983.
3 H. Rabaud and J. Clastre, *Acta Cryst.*, 1959, **12**, 911.
4 D. Hull and D.J. Bowen, *Introduction to Dislocations*, 3rd edn, Oxford: Pergamon, 1984.
5 M.D. Cohen, Z. Ludmer, J.M. Thomas and J.O. Williams, *Proc. Roy. Soc.*, 1971, A324, 459.
6 See reference 4, chapter 9.
7 J.M. Robertson, H.M.M. Shearer, G.A. Sim and D.G. Watson, *Acta Cryst.*, 1962, **15**, 1.
8 A. Klug, *Nature*, 1947, **160**, 570.
9 J.D. Wright, K. Yakushi and H. Kuroda, *Acta Cryst.*, 1978, **B34**, 1934.
10 H. Tsuchiya, F. Marumo and Y. Saito, *Acta Cryst.*, 1972, **B28**, 1935.
11 C.A. Fyfe, *J. Chem. Soc. Faraday II*, 1974, **70**, 1633, 1642.
12 C.A. Fyfe, D. Harold-Smith and J. Ripmeester, *J. Chem. Soc. Faraday II*, 1976, **72**, 2269.
13 H. Kobayashi, *Bull. Chem. Soc. Japan*, 1975, **48**, 1373.
14 G.M. Janini, K. Johnston and W.L. Zielinski Jr, *Anal. Chem.*, 1975, **47**, 670.

15 N.H. Hartshorn and A. Stuart, *Practical Optical Crystallography*, London: Arnold, 1964.

16 J.M. Thomas and J.O. Williams, *Prog. Solid State Chem.*, 1971, **6**, 119.

17 B.K. Tanner, *X-ray Diffraction Topography*, Oxford: Pergamon, 1976.

18 B.K. Tanner, *Prog. Crystal Growth Character.*, 1977, **1**, 23.

19 W. Jones and J.M. Thomas, *Prog. Solid State Chem.*, 1979, **12**, 101.

20 W. Jones, *Surface and Defect Properties of Solids*, Specialist Periodical Reports, Chemical Society, London, 1976, **5**, 65.

21 J.R. Fryer, *Mol. Cryst. Liq. Cryst.*, 1983, **96**, 275; 1986, **137**, 49.

22 T. Kobayashi and S. Isoda, *J. Mater. Chem.*, 1993, **3**, 1.

23 I. Dietrich, H. Formanek, W. von Gentzkow and E. Knapek, *Ultramicroscopy*, 1982, **9**, 75.

24 G. Binnig, H. Rohrer, C. Gerber and E. Weibel, *Phys. Rev. Lett.*, 1983, **50**, 120.

25 D. Rugar and P. Hansma, *Physics Today*, October 1990, pp. 23–30.

26 H.K. Wickramasinghe, *Sci. Am.*, October 1989, pp. 74–81.

27 G. Kaupp, *Mol. Cryst. Liq. Cryst.*, 1992, **211**, 1.

28 M. Eldrup, *Positron Annihilation*, Proc. Int. Conf., 6th., ed. P.G. Coleman, S.C. Sharma and L.M. Diana, Amsterdam: North Holland, 1982, p. 753.

29 T. Gorowek, C. Rybka and J. Wawryszczuk, *Phys. Stat. Sol. b*, 1978, **89**, 253.

30 M. Eldrup, N.J. Pedersen and J.N. Sherwood, *Phys. Rev. Lett.*, 1979, **43**, 1407.

5

Molecular motion in crystals

In chapter 3 the structures of molecular crystals were discussed largely without reference to the fact that atoms and molecules do not occupy static positions in the crystal. In fact both internal and external modes of molecular motion occur in all crystals. Internal modes include vibrations of atoms or groups within an individual molecule. External modes include motion of whole molecules, for example rigid-body torsion, libration or coupled torsion–libration (screw) motion; in-plane rotation; translational diffusion; and collective motions of molecules in crystals, known as phonons. In this chapter, the principal experimental methods used for studying these various modes of motion will be described, with examples of the type of information they can provide.

5.1 X-ray diffraction studies

A good modern X-ray diffraction study of the structure of a molecular crystal provides, in addition to information on the geometrical arrangement of atoms and molecules within the unit cell, data on the anisotropic thermal motion of the atoms in the crystal[1]. The anisotropic temperature factor is a measure of the reduction in scattering of the X-ray beam resulting from motion of each atom and is of the form

$$\exp - (\beta_{11}h^2 + \beta_{22}k^2 + \beta_{33}l^2 + 2\beta_{12}hk + 2\beta_{13}hl + 2\beta_{23}kl) \qquad (5.1)$$

where the six independent elements β_{ij} are defined with respect to the crystal axis system. By suitable transformation to orthogonal axes and determination of the eigenvalues and eigenvectors of the matrix of transformed values, the principal axes and orientation of an ellipsoid approximating to the anisotropic thermal motion of the atom may be deduced. The size of the ellipsoid is determined by the probability of the electron being inside it, which may be pre-selected, for example, at 50%. The lengths of the principal axes of the ellipsoid are proportional to the root mean square (rms) displacements of the atom in these directions. The transformations are most conveniently carried out by standard computer programs such as ORTEP[2], which will also draw these probability ellipsoids, providing a visual representation of the motion of each atom.

A useful test that these anisotropic temperature factors are truly reflecting motion of atoms rather than merely the accumulation of a variety of experimental errors (such as absorption, diffuse scattering and limited range of $\sin\theta/\lambda$ in data collection) is provided by Hirshfeld's 'rigid bond' postulate[3]. This states that the mean-square vibrational amplitudes of a pair of bonded atoms are nearly equal along the bond direction even though they may be widely different in other directions and derives from the fact that bond-stretching vibrations for atoms other than hydrogen and deuterium are normally of much smaller amplitude than other vibrations (e.g. bending, torsional and rigid-body translational and rotational oscillations). This postulate may be extended to provide a test of whether a particular group of atoms in a molecule is undergoing thermal motion as a rigid body, since for a rigid body all the atoms behave as if they were interconnected to each other. Therefore, for any pair of atoms the mean-square vibrational amplitudes along the interatomic direction must be nearly equal in a rigid body. Where this condition applies, the motion of the rigid body may be described in terms of three tensors corresponding to torsion (T), libration (L) and screw (S) motions[4]. Thus it is possible to distinguish rigid-body (external) modes of vibration from vibrations of atoms or groups within the molecule (internal modes).

In favourable cases it is even possible to estimate from the X-ray data[5] the force constants associated with these modes of vibration. The main difficulties in attempting such an analysis lie in resolving the total vibrational motion into the contributions from individual modes and in choosing appropriate potential functions to describe these modes. Where a group of atoms behaves as a rigid body mainly executing one type of motion whose nature may be deduced from chemical intuition (e.g. torsional rotation of a methyl group in a three-fold potential barrier), both of these difficulties may be overcome to a good approximation. The results of a diffraction study may then be used to deduce the moment of inertia (I) of the rigid group of atoms (from atomic coordinates) and the mean-square vibrational amplitude ($<\Phi^2>$) of the group (from thermal ellipsoids). The force constant (f) is given by

$$f = kT/<\Phi^2> \qquad (5.2)$$

or at low temperatures, where the quantum mechanical expression

$$<\Phi^2> = (h/8\pi^2 I v)\coth(hv/2kT) \tag{5.3}$$

must be used, by

$$f = 4\pi^2 I v^2. \tag{5.4}$$

Where this approach has been applied, the derived force constants and potential barriers for torsional motions of various groups of atoms are in reasonable agreement with values obtained using other experimental techniques. Although the approach is in its infancy, the amount of precise and accurate diffraction data available is increasing rapidly, and there is consequently much scope for further analysis along these lines, for example to examine the variations in torsional barriers for the same group in a variety of different crystallographic environments to distinguish intra- and intermolecular contributions.

Despite these successes, there are many situations where the interpretation of thermal ellipsoids derived from X-ray data is less straightforward. One common example is the occurrence of orientational disorder in a structure. If this is not recognised, excessively large thermal ellipsoids, often combined with abnormal molecular dimensions, are produced by application of standard anisotropic least-squares refinement of the X-ray data,

5.1. The phenanthrene site in phenanthrene/TCNB.

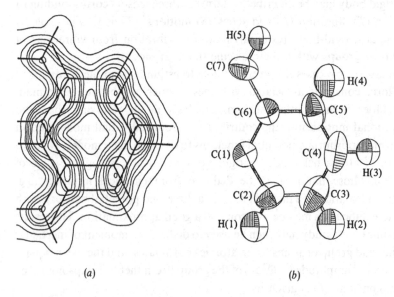

(a)   (b)

whereas a much more realistic model might be a fit of two or more constrained molecules to the lattice site with each molecule undergoing anisotropic rigid-body motion. Figure 5.1 shows one such example, the molecular complex of phenanthrene and TCNB, in which the phenanthrene molecule is disordered[6]. Figure 5.1(*a*) shows the disordered pair of orientations superimposed on an electron-density map in the plane of the molecules, while figure 5.1(*b*) shows the thermal ellipsoids for the rigid-body motion of the molecule in one of the two orientations.

In this example, the presence of two orientations is clear since they are quite distinct from each other in some regions. There are many less clear-cut cases, where two disordered orientations are sufficiently close together for it to be difficult to distinguish genuine wide-amplitude thermal motion in a flat-bottomed potential well from motion in two disordered sites (figure 5.2(*a*) and (*b*), respectively). In an extreme case, a molecule may even be executing rapid motion in a series of jumps between equivalent nuclear positions, with no detectable consequences in the diffraction data (e.g. benzene or hexamethylbenzene executing six-fold in-plane rotation). Similarly, translational diffusion in which molecules migrate through the crystal in a series of rapid jumps from one lattice site to another, facilitated by point defects, will not be revealed by X-ray diffraction studies. These limitations arise because although each diffraction event occurs in a very short space of time, a large number of such events occur in the time required for measurement of the intensity of a single diffraction maximum, so a time-average over all atomic positions is obtained. This time-averaging also means that no information is available on whether or not the motions of different molecules are correlated in any way to yield vibrational modes characteristic of the crystal lattice (i.e. phonons). Fortunately, several other experimental techniques are available for such purposes.

5.2. Potential wells for (*a*) wide-amplitude thermal motion and (*b*) disordered sites.

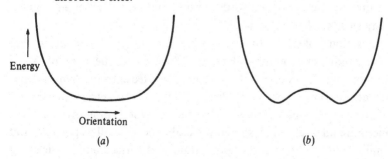

(*a*)   (*b*)

5.2   Thermodynamic studies

In principle, one of the simplest of these techniques is the calorimetric measurement of heat capacities as a function of temperature. The onset of thermal motion or disordering is associated with changes in the thermodynamic parameters of the system. The simplest classification of such transitions in thermodynamic terms is that of Ehrenfest, in which the order of a transition is the order of the lowest derivative of the Gibbs free energy which shows a discontinuity at the transition point. Thus, in first-order transitions, free energy is a continuous function of temperature, whereas the first derivatives of free energy, namely entropy and enthalpy, are discontinuous; while in a second-order transition free energy, enthalpy and entropy are all continuous functions of temperature and only the second derivatives such as specific heat show discontinuities. Although this classification has many limitations (for a more detailed discussion see reference 7) it is widely used, and thermodynamic studies of motion and disorder have been made for many molecular crystals.

A particularly simple example of such data is the quinol/HCN clathrate[8] (see chapter 2 for discussion of clathrate structures). In this compound, dipolar interactions between HCN molecules in adjacent cavities lead to an ordered low-temperature structure analogous to a ferromagnetic alloy, with all the HCN molecules aligned. At 178 K, a sharp peak in the heat capacity occurs, corresponding to the onset of thermal motion of the HCN molecules. The entropy change associated with this peak is $0.7\,R$, in good agreement with the statistical theoretical value of $R\ln2$ for the onset of reorientation between just two alternative orientations. In contrast, for larger molecules such as adamantane (chapter 2), the transition from the ordered low-temperature phase to the disordered high-temperature phase is accompanied by a much larger entropy change, since the number of available distinguishable orientations in the latter phase is large. In turn, the entropy of fusion is correspondingly low in these plastic crystals, since the degree of orientational disorder in the solid near the melting point is already close to that of the liquid.[9]

One serious obstacle to the use of thermodynamic data is that re-orientation between equivalent positions, as in the case of X-ray diffraction, is not detectable as there is no associated entropy change. Other experimental complications include the fact that calorimetric heat capacity data are average figures obtained by heating the sample over a finite temperature interval, which should preferably be as small as possible, and the influences of impurities, crystal strain and surface and particle-size effects.

5.3 Nuclear magnetic resonance studies[10]

A more versatile and widely-used technique for studying molecular motion in crystals is nuclear magnetic resonance (NMR). Three distinct types of NMR experiment may be used for this purpose, involving measurement of (i) line-widths; (ii) relaxation times; and (iii) high-resolution ^{13}C cross-polarised magic-angle spinning (CPMAS) spectra.

Line-widths
The resonance lines from a solid are generally substantially broader than those from a solution of the same material, because of magnetic dipole–dipole interactions between nuclei, or for quadrupolar nuclei, to quadrupole coupling with electric-field gradients within the crystal. The mean-square width of such a line $(\Delta B)^2$, (generally known as the *second moment*) is defined experimentally as

$$(\Delta B)^2 = \int_0^\infty (B - B_{av})^2 f(B) dB \tag{5.5}$$

where the line is of the form $y = f(B)$, B being the field and B_{av} being the field at the line centre.

If the crystal structure is known, the second moment may be calculated using a formula derived by van Vleck[11]:

$$(\Delta B)^2 = C\Sigma[(1 - 3\cos^2\theta_{jk})^2/r_{jk}^6] \tag{5.6}$$

where C is a parameter depending on gyromagnetic ratio, nuclear spin and number of nuclei in equivalent positions, r_{jk} is the length of the vector between the jth and kth nuclei, and θ_{jk} is the angle between this vector and the applied magnetic field. This is known as the rigid-lattice value. In the case of a structure in which the molecules are in motion, the summation must be carried out over all orientations in the motion. For convenience, this is often done by first calculating an intramolecular contribution, which to a good approximation is invariant with motion, and then calculating the variable intermolecular component for various models of the motion.

Comparison of the experimental data and the calculated line-widths for various models then permits characterisation of the motion. For example, for crystalline benzene[12] the experimental second moment below 90 K is 9.72 ± 0.06 G^2, in good agreement with the rigid-lattice calculated value $(9.62$ G$^2)$. Above 120 K the observed value drops to 1.6 G^2 while the calculated value assuming rotation about the six-fold axis of the molecule is 1.7 ± 0.5 G^2 (figure 5.3).

This method has been used to study the motion of the disordered donors in a series of π–π^* molecular complexes of naphthalene and pyrene[13]. X-ray diffraction data showed two donor orientations differing by an in-plane rotation of $\theta°$, while study of the temperature-dependence of the second moments revealed that the donors were jumping from one orientation to the other not just over the angle $\theta°$, with low activation barrier, but also over the larger angle $(180 - \theta)°$ with higher activation energy in several cases. In both of these examples NMR line-width data provide information not obtainable from X-ray or thermodynamic studies, although it is difficult to derive reliable activation energy barriers for the motions merely from the temperature range in which the line narrowing occurs. For this purpose, relaxation methods are preferred[14].

Relaxation times

In the NMR experiment, absorption of the radio-frequency radiation (e.g. 100 MHz) alters the proportion of nuclei in the available spin states. If the radiation is cut off, there is an exponential decay to the original thermal population of spin states, characterised by the spin–lattice relaxation time T_1 and the spin–spin (or transverse) relaxation time T_2. In a related experiment, the alignment of nuclear magnetisation with the applied field direction is transferred to a rotating field B, and the decay of the magnetisation in this low field gives the spin–lattice relaxation time $(T_{1\rho})$ in field B. This permits extension of the range of relaxation times which can be studied into a much longer time range. (A disadvantage of these NMR

5.3. Benzene.

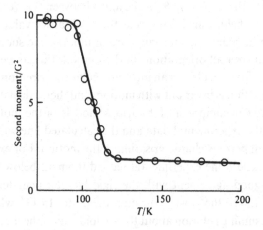

techniques is that the motion will influence the linewidth or relaxation time only if it is occurring on a time scale comparable to NMR frequencies, and thus may not necessarily correlate with the results of other techniques having different time scales.) Relaxation times T_1 and correlation (or jump) times τ_c for the motion in question are linked by the expression[15]

$$1/T_1 = KI(I+1)\gamma^4(h/2\pi)^2\Sigma r_j^{-6}\{[\tau_c/(1+\omega^2\tau_c^2)] + [4\tau_c/(1+4\omega^2\tau_c^2)]\} \quad (5.7)$$

where γ is the gyromagnetic ratio, r_j is the distance from the jth nucleus to the reference nucleus, I is the nuclear spin and K is a constant depending on the nature of the motion. The temperature dependence of T_1 can be related to the activation energy for the motion if it is assumed that

$$\tau_c = \tau_o \exp(-E/RT), \quad (5.8)$$

so that a plot of $\log T_1$ versus $1/T$ should give a V-shaped curve with a definite minimum, and limiting low-temperature and high-temperature slopes equal to E/R and $-E/R$, respectively. Where several motions occur, several such minima may be observed. Thus, in $(CH_3)_3CCl$ the plot of $\log T_1$ versus $1/T$ shows three minima, with the lowest temperature one ascribed to methyl group reorientation, the central one to molecular tumbling and the third, just below the melting point, to translational motion in the lattice (figure 5.4)[7]. A similar method may be applied for $T_{1\rho}$ data. These relaxation methods are collectively referred to as pulsed NMR methods as they involve subjecting the sample to a series of pulses of radiofrequency as opposed to the normal continuous-wave measurements of solution NMR.

CPMAS spectra

For many years solid-state workers have envied the high resolution and consequent vast potential information on molecular structure available to colleagues using solution NMR, and the development of ^{13}C CPMAS NMR spectroscopy has now made comparable resolution available for solid-state materials[16, 17]. The technique overcomes three problems hindering high-resolution solid-state NMR as follows. Broadening of the signal by ^1H–^{13}C dipolar coupling is removed by resonant ^1H decoupling; the weak signal strength consequent on long relaxation time (T_1) is improved by coherently-driven proton–carbon cross-polarisation; and broadening resulting from carbon chemical-shift anisotropy is removed by spinning the sample at rates typically 1.5–5 kHz about an axis 54.7° to the field. (The last process is at a rate fast compared with the anisotropy expressed in frequency units, and at the angle for which $(3\cos^2\theta - 1) = 0$.)

The resulting high-resolution spectra vastly increase the potential of NMR as a tool for investigating molecular motion in crystals.

In the simplest case, the use of high-resolution spectra obtained at different temperatures can reveal the onset of fluxional behaviour in which the NMR signals from chemically-inequivalent atoms are averaged by rapid reorientation. For example, pentacarbonyl(cyclooctatetraene)diiron crystals show only one ^{13}C peak for the cyclooctatetraene carbons down to $-160\,°C$, indicating that the ring is rotating rapidly even in the solid state, where the X-ray crystal structure (figure 5.5) would predict inequivalent ring carbons. Further cooling to 32 K leads to a splitting of this peak into two, with intensity ratios 1 : 3 for carbons labelled 1 and 2, 3, 4, respectively, in figure 5.5, showing that the motion slows down at these very low temperatures.

A more important advantage of the CPMAS technique is that, because of the high resolution, it is possible to study the relaxation times and linewidths for individual atom resonances. In the proton-relaxation techniques discussed above, spin diffusion results in averaging of relaxation

5.4.

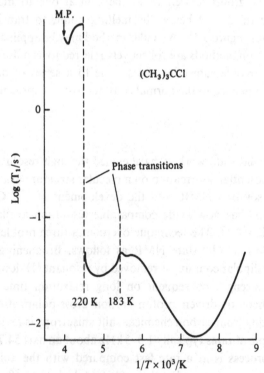

behaviour over all protons in the sample to give a single relaxation time at a particular temperature. This makes it difficult to interpret the data unambiguously in terms of particular motional modes. Identification of a particular resonance in a high-resolution CPMAS spectrum, followed by determination of the associated relaxation times and linewidth, permits a more positive identification of the particular motion contributing to the relaxation or linewidth. For example, measurements of linewidth as a function of temperature for the methyl carbons in hexamethylbenzene and hexamethylethane have yielded values for the methyl and hexad rotational barriers, respectively, in good agreement with values derived from conventional proton T_1 and second-moment data, while indicating clearly which atoms are being affected by the motion. Thus, the quaternary carbons in hexamethylethane give linewidths very little influenced by temperature, while the methyl carbon linewidths are strongly temperature-dependent.

Although these techniques are currently in their infancy, it is to be expected that in future they will contribute extensively to our knowledge of motion in molecular crystals, particularly in view of the rapid increase in availability of NMR spectrometers having variable-temperature CPMAS facilities.

5.4 Nuclear quadrupole resonance studies

Another nuclear-resonance technique for the study of molecular motion is nuclear quadrupole resonance (NQR). Nuclei with spin greater than half have quadrupole moments arising from non-spherical charge distribution. In an inhomogeneous electric field (i.e. in a field gradient), the energy (quantised) depends on the orientation of the quadrupole with

5.5.

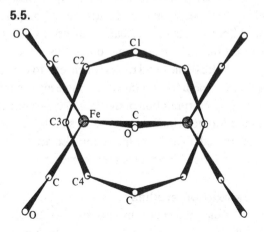

respect to the field gradient. Transitions between different orientations may be induced by radio-frequency radiation, producing the quadrupole resonance spectrum with resonance frequencies proportional to the electric field gradient at the nucleus and to the nuclear quadrupole moment. The motion of molecules in crystals frequently alters the electric-field gradients in the vicinity of different nuclei, and, since this change is generally a reduction in field gradient resulting from the averaging effect of the motions, a corresponding decrease in the NQR frequency is observed. For example, the onset of molecular motion in solid dichloromethane has been studied by measuring the reduction in NQR frequency for the chlorine nuclei as the sample temperature increases[18]. Assuming that the field gradient at the chlorine nucleus has axial symmetry about the C–Cl bond, the resonance frequency at temperature T (v_T) is related to that for the stationary molecule (v_Q) by

$$v_T = v_Q(1 - \tfrac{3}{2} < \theta_x^2 >)$$ (5.9)

where $< \theta_x^2 >$ is the mean-square angular displacement of the C–Cl bond from its equilibrium position. Since

$$< \theta_x^2 > = (h/4\pi^2 I_x v_x)\{0.5 + [\exp(h v_x/kT) - 1]^{-1}\}$$ (5.10)

where I_x = moment of inertia about the specified axis x,
and v_x = corresponding vibrational frequency for the oscillation,

$$[(v_T - v_0)/v_Q] = -(3h/8\pi^2 I_x v_x)[\exp(h v_x/kT) - 1]^{-1}$$ (5.11)

where v_0 is the limiting low-temperature NQR frequency (often approximated to equal v_Q). A fit of this equation to data on $(v_T - v_0)/v_Q$ as a function of temperature yields values of v_x and I_x (figure 5.6). Comparing the moment of inertia with known data on molecular dimensions then permits identification of the direction of the axis of rotation, so this method reveals both the frequency and nature of the oscillations which lead to reduction in the NQR frequency. The method is limited to molecules containing quadrupolar nuclei and the interpretation of the data may become difficult if several different modes of molecular motion occur simultaneously over the temperature range studied. (The latter limitation also applies to many of the other techniques described in this chapter, however.)

5.5   Dielectric relaxation studies

The motion of polar molecules in crystals may be studied by dielectric measurements. The dielectric constant (ε) of a medium is related to the polarisability (α) and dipole moment (μ) of the molecules via the Debye equation:

$$(\varepsilon-1)/(\varepsilon+2)=(N/3)\{\alpha+(\mu^2/3kT\varepsilon_0)\}. \tag{5.12}$$

In order for the second term on the right-hand side of this equation to contribute to the dielectric constant, the polar molecules must be able to rotate so that the dipole remains aligned with the electric field. This occurs readily in liquids but generally not in solids, so that the dielectric constant of many substances decreases sharply on solidification (e.g. that of nitromethane drops by a factor of 12 at the melting point ($-28.5\,°C$) when measured with a 70 kHz field). However, in the case of polar molecules capable of forming plastic crystal phases (e.g. *t*-butyl chloride) the rapid re-orientation of the molecules on their lattice sites leads to almost no change in the dielectric constant on freezing. On further cooling, phase changes occur in which the re-orientation stops, leading to a reduction in the dielectric constant.

Although useful information on gross changes in molecular motion is available from such studies of the temperature-dependence of the dielectric constant, more-detailed information can be obtained using the technique of dielectric relaxation, in which dielectric constant is measured over a wide range of frequencies of the applied field. At low frequencies, molecular

5.6. Temperature-dependence of NQR frequency for CH_2Cl_2.

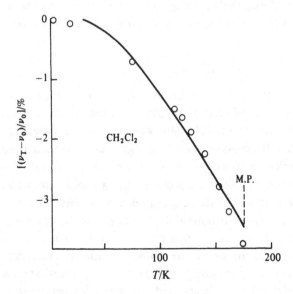

re-orientation occurs fast enough to follow the reversals of the applied field, but at high frequencies the motion is too slow to follow the field and there is a phase lag between the polarisation and the applied field. This leads to energy transfer from the applied field to the medium in a range of frequencies (ω) comparable to $\frac{1}{\tau}$, where τ is a relaxation time characterising the re-orientational motion. This energy transfer is known as dielectric loss, and determination of the frequency at which it is maximised yields the relaxation time for the motion invoiced. In the region of frequencies where dielectric loss occurs, the dielectric constant is a complex quantity

$$\varepsilon = \varepsilon' - i\varepsilon'' \tag{5.13}$$

where ε' is the measured dielectric constant and ε'' is the dielectric loss factor.

Debye showed that

$$\varepsilon = \varepsilon_\infty + [(\varepsilon_0 - \varepsilon_\infty)/(1 + i\omega\tau)] \tag{5.14}$$

where ε_0 and ε_∞ are the dielectric constants at zero and infinite frequency, respectively. Combination of these two equations yields

$$\varepsilon' = \varepsilon_\infty + [(\varepsilon_0 - \varepsilon_\infty)/(1 + \omega^2\tau^2)] \tag{5.15}$$

and

$$\varepsilon'' = [(\varepsilon_0 - \varepsilon_\infty)\omega\tau]/(1 + \omega^2\tau^2) \tag{5.16}$$

(see figure 5.7). Hence, the maximum value of ε'' occurs when $\omega\tau = 1$, and $\varepsilon''_{max} = (\varepsilon_0 - \varepsilon_\infty)/2$. Combining the equations for ε' and ε'' to eliminate $\omega\tau$ yields

$$\varepsilon''^2 + [\varepsilon' - (\varepsilon_0 + \varepsilon_\infty)/2]^2 = (\varepsilon_0 - \varepsilon_\infty)^2/4, \tag{5.17}$$

which is of the form $x^2 + y^2 = r^2$, i.e. the equation of a circle.

Since only positive values of ε'' are physically meaningful, this relationship corresponds to a semicircular plot, known as a Cole–Cole plot, if ε'' is plotted against ε'. If the assumption of a single relaxation time is valid, the plot is a complete semicircle with centre on the ε'-axis, but if several relaxation times are involved the plot becomes a smaller arc, with centre below the ε'-axis. The wider the range of relaxation times, the lower is the centre of the arc. If the distribution of relaxation times is not symmetrical, the plots become skewed rather than symmetrical arcs. These variations[19] as well as full derivations of the above equations[20] and details of experimental methods[20] are discussed fully in more specialised reviews and texts. Clearly, if ε' and ε'' are determined over a wide frequency range, plots of ε' versus ε'' reveal the nature of the distribution of relaxation times, while determination of the frequency of maximum dielectric loss ε'' yields the value of the principal relaxation time. Temperature-dependence of these quantities further reveals the energy barriers associated with the relevant motions, while studies of the dependence of the dielectric loss on orientation of single-crystal samples with respect to the field permit identification of the direction of the axis of rotation in favourable cases.

An elegant example is provided by the studies of rotation of methanol and acetonitrile molecules in quinol clathrates[21]. The CH_3CN molecule is a tight fit in the cavity of the quinol lattice, and the c-axis (which is also the long axis of the CH_3CN molecule) is elongated from its normal value as a result, whereas the CH_3OH molecule is more easily accommodated. These differences are reflected in the dielectric relaxation times at $19\,°C$, which are $2 \times 10^{-12}\,s$ for the CH_3OH compound and $5 \times 10^{-5}\,s$ for the CH_3CN. Studies of the temperature-dependence of these relaxation times show that the slower more restricted rotation of CH_3CN has an associated activation energy of $75.3\,kJ\,mol^{-1}$, whereas that for CH_3OH rotation is only $9.6\,kJ\,mol^{-1}$. Measurements of the dielectric loss for the CH_3CN compound for samples with the field parallel and perpendicular to the c-axis are shown in figure 5.8, confirming that the CH_3CN molecules are aligned along the c-axis and that molecular motion involves rotation about axes perpendicular to the c-axes (i.e. so that the dipoles follow field oscillations parallel to the c-axis).

The principal disadvantage of this technique is its dependence on the

5.7.

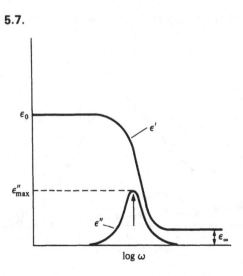

motion of polar groups. The requirement of a dipole moment not only excludes the possibility of studying motions of non-polar molecules, but also means that it is impossible to detect motions of certain groups within polar molecules (e.g. methyl group rotation in the above examples of clathrated CH_3OH and CH_3CN). Furthermore, the link between the observed relaxation times and specific modes of motion may be difficult to establish conclusively, particularly for cases having wide and asymmetric distributions of relaxation times.

5.6   Collective motions of molecules in crystals
Internal vibrational modes of free molecules have been widely studied by infra-red and Raman spectroscopy. These techniques may also be applied to molecular crystals to yield information on the collective motions of molecules in the crystal as well as on internal vibrational modes of the molecules.

The latter bands are modified in several respects compared with the corresponding free-molecule bands. Most obviously, since molecular rotation is generally restricted in crystals compared with fluid phases, rotational fine structure is frequently lost and the spectral lines are therefore narrower. Other effects include frequency shifts, splitting of bands into several components, removal of degeneracy present in the free molecule, relaxation of selection rules, and dichroism arising from the specific orientations of molecules in the crystal[22].

In general, however, these internal vibrational modes are only weakly

5.8. Dielectric loss for CH_3CN in quinol clathrate.

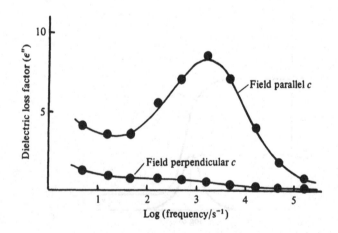

perturbed by the crystal lattice and their dependence on temperature and crystal phase is therefore also slight. The collective motions of molecules in crystals which lead to the quantised transfer of vibrational energy through the lattice as phonons can be studied only partially using infra-red and Raman spectroscopy and require the introduction of several new concepts from solid-state physics for their description[23]. They are classified in two main ways, namely as longitudinal or transverse (depending on whether the principal motion of the individual molecules is parallel or perpendicular to the direction of propagation of the wave in the crystal), and as acoustic or optical (depending on whether nearest-neighbour molecules move in the same general direction or in opposite directions in the vibration). Figure 5.9 illustrates the differences between transverse acoustic and optical modes.

Since the relative mutual displacements of nearest-neighbour molecules *A* and *B* are larger for optical than for acoustic modes, optical modes tend to have higher frequencies than acoustic modes. Classically, for longitudinal waves the propagation velocity v is determined by the adiabatic elastic bulk modulus or 'stiffness coefficient' (B_s) by the relationship:

$$v = \lambda v = \sqrt{(B_s/\rho)} \tag{5.18}$$

where λ is the wavelength, v the frequency and ρ the density. Therefore, for an ultrasonic wave of frequency 1 GHz travelling at 10^6 cm s^{-1}, the wavelength is 10^{-3} cm. This is much larger than the intermolecular distance. Similarly, a light wave travelling at 3×10^{10} cm s^{-1} with frequency 3×10^{12} Hz ($\equiv 100$ cm^{-1}) has a wavelength of 10^{-2} cm, which is again much greater than intermolecular spacings. However, if instead of light or sound

5.9.

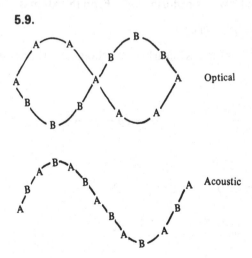

Optical

Acoustic

a neutron beam is used this condition is no longer valid, since

$$\lambda = h/mv = 3.956 \times 10^{-7}/v \qquad (5.19)$$

and frequency

$$v = v/\lambda = v^2/(3.956 \times 10^{-7}) \qquad (5.20)$$

so that for $v = 3 \times 10^{12}$ Hz, $v = 1089$ m s^{-1} and $\lambda = 3.6 \times 10^{-10}$ m, which is of the same order as the lattice spacing. Under these conditions, it is no longer correct to assume that the wave velocity is constant independent of wavelength, and the relationship between velocity and wavelength will now be derived for the simple case of a linear one-dimensional lattice. Since the dependence of velocity on wavelength is responsible for the dispersion of light by a prism, the relationship is known as a phonon dispersion curve and is normally expressed as the plot of phonon frequency, ω, as a function of wave-vector k (where $|k| = 1/\lambda$).

Consider a one-dimensional lattice at equilibrium (figure 5.10(a)) and during the passage of a longitudinal wave (figure 5.10(b)). The lattice constant is a, the atoms are labelled $r-2, r-1, r, r+1, r+2$, etc. and the associated displacements are $u_{r-2}, u_{r-1}, u_r, u_{r+1}, u_{r+2}$, etc. For a wave of amplitude A, wave-vector k and frequency ω, the displacement of the rth atom is given by

$$u_r = A \exp[i(kra - \omega t)]. \qquad (5.21)$$

Thus $d^2u_r/dt^2 = -\omega^2 A \exp[i(kra - \omega t)]$
$$= -\omega^2 u_r. \qquad (5.22)$$

From Newton's second law, the restoring force F_r on this atom is given by

$$F_r = m(d^2u_r/dt^2) = -m\omega^2 u_r. \qquad (5.23)$$

If we now assume that this force is linearly proportional to the extension or

5.10.

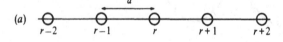

(a)

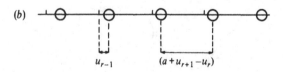

(b)

compression of the nearest-neighbour distance (Hooke's law), with proportionality constant p,

$$F_r = p(u_{r+1} - u_r) - p(u_r - u_{r-1})$$
$$= p(u_{r+1} + u_{r-1} - 2u_r). \tag{5.24}$$

Thus
$$\omega^2 = (p/m)[2 - (u_{r+1}/u_r) - (u_{r-1}/u_r)]$$
$$= (p/m)[2 - \exp(ika) - \exp(-ika)]$$
$$= 2(p/m)[1 - \cos(ka)]$$
$$= 4(p/m)\sin^2(ka/2) \tag{5.25}$$

i.e.
$$\omega = \pm 2[\sqrt{(p/m)}]\sin(ka/2) \tag{5.26}$$

(see figure 5.11). In the classical case where $\lambda \gg a$ so that $ka \ll 1$, $\sin(ka/2) \to ka/2$ and

$$\omega/k \approx a\sqrt{(p/m)} = v. \tag{5.27}$$

Naturally, the assumptions used in the above treatment are of limited validity for molecular crystals, so the detailed form of the actual phonon dispersion curve[24] will not be identical to the sine curve of figure 5.11. (The short-range nature of the forces in molecular crystals implies that considering only nearest neighbours may be a good approximation, but the assumption of a linear relationship between force and displacement will be less valid.) However, it is clear from this derivation that the study of phonon dispersion curves can provide valuable information on the form of the intermolecular potential in the crystal, particularly when carried out over a range of temperatures and pressures. Equally clearly, it is apparent from the earlier considerations above that studies of lattice vibrations with light or sound waves of appropriate frequency will involve wavelengths which are

5.11.

much larger than the lattice spacing, so that only the frequencies close to zero wave-vector will be probed. For these reasons, inelastic neutron scattering is the preferred technique for studying phonon dispersion curves[25]. Nevertheless, since the low-frequency lattice vibrations are strongly influenced by crystal structure and temperature, their study using conventional infra-red and Raman spectroscopy is important in the investigation of phase transitions in molecular crystals. For example, the α, β and γ phases of p-dichlorobenzene each have characteristic low-frequency Raman spectra which have been used to follow phase transitions[26].

5.7   Neutron-scattering studies

Since neutrons have wavelengths comparable to intermolecular distances and energies comparable to phonon energies, they can exchange energy quanta with lattice vibrations and undergo inelastic scattering. Experimentally, this is studied using a triple-axis neutron spectrometer as shown schematically in figure 5.12. The first monochromator selects neutrons of a particular wavelength, which are collimated in a particular direction defined by $2\theta_M$. These neutrons, of wave-vector k_0, are inelastically scattered by the sample, and the wavelengths of the scattered neutrons emerging at angles $2\theta_s$ are determined by the analyser by finding the Bragg angles $2\theta_A$ at which peaks occur. Thus, the magnitude and direction of the scattered wave-vector k_1 are determined. Hence the phonon wave-vector $k = k_1 - k_0$ and the phonon frequency is given by

$$\hbar\omega = (\hbar^2/2m)(k_0^2 - k_1^2) \tag{5.28}$$

where m is the neutron mass.

In addition to determining full phonon dispersion curves, this versatile technique may be used in a way analogous to Raman spectroscopy (but without the limitations of the selection rules applying to the latter technique) to study, for example, torsional oscillation frequencies. Furthermore, neutrons, unlike photons, have considerable momentum which can be transferred to a crystal lattice during inelastic scattering; this is revealed as a dependence of the intensity of peaks in the neutron-scattering spectrum on the scattering angle, from which it is possible to determine the amplitude of the vibration or motion under study. Further information on molecular motion is available from quasi-elastic scattering of neutrons, in which normal elastic-scattering peaks are broadened by the motion of the scattering molecules via a Doppler effect. For example, neopentane undergoes a phase transition at 140 K from an ordered phase to an orientationally disordered plastic crystal phase, which is reflected in a reduction in the intensity of the sharp elastic-scattering peaks in the neutron spectrum accompanied by the appearance of a broad component associated with the re-orientational motion[27].

Despite the versatility of neutron-scattering techniques for the study of molecular motion, they suffer from the disadvantage that neutron experiments can be carried out only at centralised facilities since, ideally, dedicated nuclear reactors are involved.

5.8   Diffusion in molecular crystals

All of the above techniques have been discussed from the point of view of characterising individual or collective motions of molecules which remain associated with individual lattice sites. However, crystals composed of certain types of molecule, in which both motion and point-defect formation are facile, display molecular motion over larger distances, namely translational diffusion from one lattice site to the next. Such diffusion is largely limited to plastic crystals (see chapter 3) and rare-gas solids, although diffusion rates have also been measured for a range of aromatic hydrocarbons and related materials (e.g. biphenyl, benzoic acid,

5.12. Triple-axis neutron spectrometer.

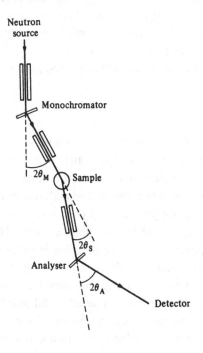

imidazole)[28]. Although the NMR methods described earlier may be used to study diffusional jump motion, a more direct technique is to study the diffusion of a radioactively-labelled tracer molecule into the lattice of chemically identical non-labelled molecules.

In this method, a thin layer of the material is deposited onto one face of the crystal, which is then annealed at a constant temperature for a fixed time, following which the sample is sectioned using a microtome and the activity in each section is determined to deduce the rate of diffusion. Great care is needed both in sample preparation and in interpretation of the data if significant results are to be obtained[29]. Diffusion occurs more rapidly along dislocations and grain boundaries than through the bulk lattice, so such defects should be avoided by careful purification and crystal growth as well as by careful handling of the sample to avoid strain-induced defects. The choice of labelling isotope should also be made with care, to avoid radiation-induced defects which could assist diffusion. Even with full attention to these points, some defect diffusion may still occur, and a careful analysis of the tracer distribution profile in the sample following the diffusion annealing process may permit isolation of bulk and defect components of the total process. Diffusion measurements have also been made by following the plastic deformation of suitable molecular crystals under conditions of low applied stress and high temperature. Although such experiments are simple in principle, they are difficult in practice and great care must be taken to avoid exceeding the applied stress under which genuine steady-state deformation ('creep') occurs.

Experimental diffusion-coefficient data are usually presented in the form of an Arrhenius expression

$$D = D_0 \exp(-Q/kT) \tag{5.29}$$

where D is the diffusion coefficient ($= <R^2>/6t$ where $<R^2>$ is the mean-square displacement of a molecule in time t), and $Q = h + \Delta h$, where h is the enthalpy of vacancy formation and Δh is the barrier enthalpy associated with the jump from a lattice site to a neighbouring vacancy. Table 5.1 shows values of D, D_0 and Q for a range of compounds, together with the ratio of Q/L_s (where L_s = latent heat of sublimation)[28]. The small range of values for Q/L_s supports the view that diffusion involves vacancy formation and migration, and a more detailed comparison of theory and experiment supports this conclusion. Calculations using the atom–atom potential method outlined in chapter 2 have proved useful in estimating barrier heights, not only for diffusion processes but also for many other molecular motions which have been studied by the techniques described above. These experimental and theoretical methods permit fairly complete characterisation of the dynamic aspects of molecular crystals, which are of particular relevance to optical and electrical properties of these materials, as will be shown in chapters 6 and 8.

Table 5.1. *Diffusion coefficients and related data for some molecular crystals (c.f. equation 5.29)*

Compound	$D/m^2 s^{-1}$ at MP	$D_0/m^2 s^{-1}$	$Q/kJ mol^{-1}$	Q/L_s
Benzene	1×10^{-13}	1.4×10^5	96.0	2.1
Naphthalene	1×10^{-15}	2×10^{11}	178.6	2.5
Anthracene	3×10^{-16}	1×10^6	202	2.1
Phenanthrene	1.7×10^{-15}	3×10^{13}	202	2.3
Biphenyl	7×10^{-16}	2.5×10^{10}	168	2.3
Cyclohexane*	6×10^{-13}	1×10^5	92	2.0
Hexamethylethane*	2.4×10^{-12}	2.2	86	2.2
Adamantane*	7.6×10^{-14}	1.6	139	2.1
Norbornylene*	1.6×10^{-13}	3×10^{-5}	49	1.5

*Denotes plastic crystal

References

1 B.T.M. Willis and A.W. Pryor, *Thermal Vibrations in Crystallography*, Cambridge: Cambridge University Press, 1975.

2 C.K. Johnson, *ORTEP: a FORTRAN Thermal Ellipsoid Plot Program*, ORNL-3794, Oak Ridge National Laboratory, Tennessee.

3 F.L. Hirshfeld, *Acta Cryst.*, 1976, **A32**, 239.

4 V. Schomaker and K.N. Trueblood, *Acta Cryst.* 1968, **B24**, 63.

5 K.N. Trueblood and J.D. Dunitz, *Acta Cryst.* 1983, **B39**, 120.

6 J.D. Wright, K. Yakushi and H. Kuroda, *Acta Cryst.* 1978, **B34**, 1934.

7 N.G. Parsonage and L.A.K. Staveley, *Disorder in Crystals*, Oxford: Oxford University Press, 1978.

8 T. Matsuo, H. Suga and S. Seki, *J. Phys. Soc. Japan*, 1968, **25**, 641.

9 J.N. Sherwood (ed.), *The Plastically Crystalline State*, London: Wiley, 1979.

10 C.A. Fyfe, *Solid-State NMR for Chemists*, Guelph: CFC Press, 1983.

11 J.H. van Vleck, *Phys. Rev.* 1948, **74**, 1168.

12 E.R. Andrew, *J. Chem. Phys.* 1950, **18**, 607. E.R. Andrew and R.G. Eades, *Proc. Roy. Soc.* 1953, **218A**, 537.

13 C.A. Fyfe, *J. Chem. Soc. Faraday II*, 1974, **70**, 1633, 1642.

14 See, for example, C.A. Fyfe, D. Harold-Smith and J. Ripmeester, *J. Chem. Soc. Faraday II*, 1976, **72**, 2269–2282.

15 N. Bloembergen, E.M. Purcell and R.V. Pound, *Phys. Rev.* 1968, **73**, 679.

16 C.S. Yannoni, *Acc. Chem. Res.* 1982, **15**, 201.

17 J.R. Lyerla, C.S. Yannoni and C.A. Fyfe, *Acc. Chem. Res.* 1982, **15**, 208.

18 H.S. Gutowsky and D.W. McCall, *J. Chem. Phys.* 1960, **32**, 548.

19 R.H. Cole, in *Physics of Dielectric Solids*, ed. C.H.L. Goodman, Institute of Physics Conference Series No. 58, 1980, p. 1.

20 N.E. Hill, W.E. Vaughan, A.H. Price and M. Davies, *Dielectric Properties and Molecular Behaviour*, Amsterdam: van Nostrand Reinhold, 1969.

21 J.S. Dryden, *Trans. Faraday Soc.* 1953, **49**, 1333.

22 D.A. Dows, in *Physics and Chemistry of the Organic Solid State*, Vol. 1, Ch. 11, ed. D. Fox, M.M. Labes and A. Weissberger, New York: Interscience, 1963.

23 See, for example, J.S. Blakemore, *Solid State Physics*, Ch. 2, Philadelphia: Saunders, 1969.

24 W. Cochran, *The Dynamics of Atoms in Crystals*, Ch. 5, London: Arnold, 1973.

25 See, for example, U. Schmelzer, E.L. Bokhenkov, B. Dorner, J. Kalus, G.A. Mackenzie, I. Natkaniec, G.S. Pawley and E.F. Sheka, *J. Phys. C.*, 1981, **14**, 1025.

26 M. Ghelfenstein and H. Szwarc, *Mol. Cryst. Liq. Cryst.* 1971, **14**, 283.

27 U. Dahlborg, C. Gräslund and K.E. Larsson, *Physica*, 1972, **59**, 672.

28 A.V. Chadwick, in *Mass Transport in Solids*, NATO Advanced Study Institute Series B, **97**, Ch. 11, ed. F. Bénière and C.R.A. Catlow, New York: Plenum, 1983.

29 J.N. Sherwood, in *Surface and Defect Properties of Solids*, Vol. 2, ed. M.W. Roberts and J.M. Thomas, London: Chemical Society, 1973, p. 250.

6

Optical properties of molecular crystals

Although the forces between molecules in crystals are weak and short-range, and the overlap between the orbitals of adjacent molecules in the lattice is small, there are substantial differences between the electronic spectra of molecular crystals and free molecules. Some of these differences arise from interactions between the electronic states of a molecule and those of molecules in the immediate vicinity, while others arise as a consequence of the collective properties of the crystal lattice. Similarly, there are differences between the vibrational spectra of solids and free molecules, as mentioned in chapter 5, of which some may be regarded as effects resulting from changes in the local environment of a molecule or group while others, such as phonon dispersion curves, are characteristic of the lattice as a whole. This sensitivity of optical properties to the structure of, and interactions within, molecular crystals implies that studies of the spectra of molecular crystals can yield a large amount of information on these structures and interactions. Furthermore, spectroscopic studies of single crystals of known structure and orientation using polarised light provide data on the directions of the transition dipole moments relative to molecular axes which can greatly assist the assignment and theoretical interpretation of the spectra.

6.1   Methods for measuring absorption spectra of solids

Experimentally, measurements of absorption spectra of solids present a number of practical difficulties, particularly in the case of electronic spectra. These arise principally as a result of the high concentration of matter in solids, which leads to very strong optical absorbance for all but very thin samples. For example, lead phthalocyanine (density $1.95 \, \text{g cm}^{-3}$, rmm 719.7) has a π–π^* transition leading to an absorption maximum at

680 nm in solution with extinction coefficient $1.4 \times 10^5 \, \mathrm{l \, mol^{-1} \, cm^{-1}}$. If the transition intensity is of similar magnitude in the solid state, the sample thickness for which the absorbance would reach 1.5 may be calculated, using the Beer–Lambert law, to be

$$1.5 \times (719.7/1950)/(1.4 \times 10^5) = 4 \times 10^{-6} \, \mathrm{cm}.$$

Since stray light in monochromators imposes a limit on the lowest significantly measurable intensity of transmitted light, it is generally desirable to work with absorbances no higher than this, so very thin samples are needed if transmission spectra are to be measured.

In the infra-red region, this problem is generally overcome by using fine powder in a nujol mull or KBr matrix, but these methods are much less satisfactory for visible/ultra-violet spectra, because of the much higher light scattering when shorter wavelengths are used. (Scattering is proportional to λ^{-4}.) Although the methods may be used with limited success if a non-absorbing sample of similar scattering power (e.g. a pure KBr pellet or a piece of filter paper made translucent by soaking in nujol or a similar liquid) is placed in the reference beam, the resulting loss of signal intensity increases noise and reduces resolution. Diffuse-reflectance spectroscopy and single-crystal transmission and reflectance spectroscopy are preferred methods.

Perfectly smooth surfaces give specular reflectance of a light beam, with angle of incidence equal to angle of reflection, whereas rough surfaces give diffuse reflectance with the reflected beam diffused over a wide angular range. Diffuse-reflectance attachments are available for many commercial spectrophotometers, permitting measurement of the reflectivity ($R_\infty =$ % reflectance/100). The reflectivity depends on both the absorption coefficient (K) and the scattering coefficient (S) via a relationship known as the Kubelka–Munk function[1]

$$K/S = (1 - R_\infty)^2/2R_\infty. \tag{6.1}$$

Two problems in using this method are the difficulty of finding a 100%-reflecting reference standard and the fact that the scattering coefficient S is generally not known. The former problem may be overcome by using secondary standards of known reflectivity (e.g. freshly-prepared MgO with reflectivity close to 98% across the visible range of wavelengths), while the latter problem may be avoided by use of hand-ground particles, which are generally sufficiently large for S to be independent of wavelength across the visible region so that K/S is proportional to K only. Very intensely absorbing materials may be diluted with inert substances which have no optical absorbance in the range of interest (e.g. KBr) to produce very thin layers of the material coating the surfaces of particles of the inert diluent. Further practical and theoretical aspects of this cheap and effective technique are comprehensively discussed in reference 1. One important disadvantage of this technique is that it cannot provide information on the polarisation of the transitions studied, since the sample is a powder.

Single-crystal transmission spectra, which do provide such information, require very thin crystals as already explained. Such crystals are also usually small in all their dimensions, and this requires the use of microspectrophotometers. In a microspectrophotometer, a pinhole is illuminated and a reflecting microscope objective is used to project a much-reduced image of the pinhole onto a selected area of the very small crystal under study. The transmitted light is focussed by a second reflecting condenser objective onto a second pinhole, which serves to reduce stray light, before reaching the photomultiplier detector. Although this is simple in principle, precision mechanical and optical components are required and good commercial microspectrophotometers are therefore expensive and available only in a small number of laboratories. When fitted with polarising prisms and rotating sample stages, they permit measurement of polarised spectra, but difficulties can occur with very intense transitions, for which the extremely thin samples necessary may be too small even for a microspectrophotometer. Also, the space available in the vicinity of the sample is very limited, so that measurements over a range of temperature are not usually possible.

These difficulties are avoided in single-crystal reflectance spectroscopy[2], in which specular reflectance is measured using well-formed crystal faces illuminated by the projected image of a pinhole. Here the source and detector are on the same side of the sample, which may therefore be mounted on a heated or cooled stage with no space problems. The method relies on the fact that there is a relationship between absorption and reflectivity, so that reflectivity data can be transformed (using a process known as the Kramers–Kronig transformation) to yield the absorption spectrum[2,3]. Transmission and reflectance spectra of single crystals are complementary techniques, since the former is at its best for relatively weak transitions, whereas the latter is particularly useful for the study of intense transitions.

Many of the difficulties associated with the above techniques arise because in normal spectroscopy absorbed light is not measured directly but inferred from the reduction in intensity of light transmitted or reflected by the sample. These difficulties would be overcome if the amount of light absorbed could be measured directly. This has recently become possible with the development of photo-acoustic spectroscopy[4-6] (PAS), which depends on the fact that non-radiative decay of excited states leads to generation of heat in the sample. Hence, illumination of the sample with chopped light gives a thermal wave with frequency equal to the chopping frequency, leading to a periodic change in the surface temperature of the sample. If the sample is in a sealed chamber filled with an appropriate gas (air, helium, etc.) these surface-temperature fluctuations cause corresponding expansion and contraction of the gas. Acoustic waves are thus produced, of frequency equal to the chopper frequency and amplitude determined by a range of factors including the absorption coefficient (α) and thermal diffusivity (x) of the sample. These waves are detected by a microphone in the sealed chamber.

The absorption of light is characterised by the optical absorption length (l), where $l = 1/\alpha$, while the transfer of the resulting heat to the surface is governed by the thermal diffusion length (μ_s), where $\mu_s = \sqrt{(2x/\omega)}$, and ω is the chopping frequency. If the sample absorbs very strongly and is thick, the optical absorption length is less than the thermal diffusion length, so all the incident light is absorbed and contributes to the signal. In these conditions the signal is proportional to the incident light intensity and not to the absorption coefficient, and *photo-acoustic saturation* occurs. This may be avoided either by using a sample of thickness less than the optical absorption length (e.g. by diluting as in diffuse-reflectance spectroscopy) or by increasing the chopper frequency to reduce the thermal diffusion length (if $l > \mu_s$, the signal is proportional to the amount of radiation absorbed within a distance μ_s of the surface, which is dependent only on the absorption coefficient in these conditions). Figure 6.1 illustrates these effects of reducing sample thickness or increasing modulation frequency.

By varying the modulation frequency it is thus possible to observe the absorption spectra of layers of varying thickness and hence check for uniformity of the sample or probe multi-layer structures. The quality of the sample surface and scattering of light by the sample do not greatly influence the quality of the spectra obtained, in contrast to the situations for specular reflectance or transmission spectra, respectively. Furthermore, since non-radiative decay processes are required for production of heat subsequent to light absorption, the technique provides a direct measure of the probability of such processes compared with other fates of excited states (e.g. fluorescence) which do not lead to generation of heat within the sample. Experimentally, signals from the sample and from a black-body reference are compared, after amplification, using a lock-in amplifier to select signals with the same frequency as the modulation and at a particular phase with respect to the modulation cycle. Although care is needed to ensure good vibration-free acoustic isolation of the sample cell and detectors, this technique has great potential for molecular crystal samples and deserves wider application.

The problems associated with intense absorbance are much smaller if the spectra of dopants, present as guests in low concentration in a host crystal lattice, are studied. In the ideal case of a guest molecule which fits perfectly onto a host-lattice site with no distortion of the surrounding lattice, measurement of polarised absorption spectra will reveal the direction of the transition moment of the guest relative to the host lattice. If the guest occupies several different sites in which the molecular environment is different, several separate but overlapping absorption-band systems may be observed, differing in intensity, polarisation and structure, and information on these sites may be deduced by comparison of observed spectra with predictions of models[7]. Studies of the fluorescence spectra of doped molecular crystals are, however, of wider interest than absorption spectra, since it is observed that for impurities with lower-lying orbitals than those of the host molecule, excitation of the host leads to fluorescent emission characteristic of the guest in many cases. Thus, for example, as early as 1934

6.1 Influence of (*a*) sample thickness and (*b*) chopper frequency on photo-acoustic spectra. (Re-drawn after Perkampus[6].)

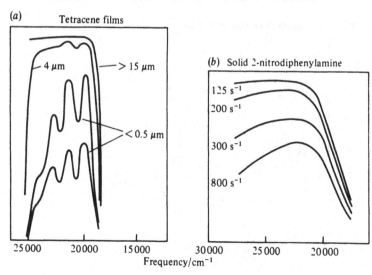

the green fluorescence of an anthracene crystal was explained as arising from emission from traces of tetracene impurities[8]. This implies energy transfer through the crystal, the origin of which will be discussed later in this chapter. In the study of such processes, fluorescence lifetimes and the rate of energy transport through the crystal must be measured, and many elegant experiments of considerable technical complexity have been carried out to this end. These have been made possible by the advent of lasers (which provide very fast, high-intensity pulses of monochromatic light) together with modern electronics (e.g. for time-resolved single-photon counting) and low-temperature techniques.

6.2   Differences between crystal, solution and gas-phase spectra

The earliest studies of the spectra of organic molecular crystals were reported in 1926 by Pringsheim and Kronenberger[9], who emphasised the similarities between crystal, solution and gas-phase spectra of species such as benzene. For example, figure 6.2 shows the spectra of an anthracene crystal and a solution of anthracene in ethanol, at 90 K[10]. This general similarity is expected according to the simplest theoretical model of molecular crystals – the *oriented-gas model*. Interactions between molecules are neglected in this model, and the crystal lattice is treated simply as a system for holding the individual molecules in specific orientations. Closer examination of spectra such as those in figure 6.2 shows differences in the

6.2. Anthracene absorption spectra.

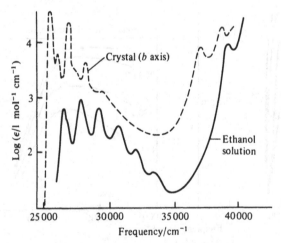

positions, shapes and intensities of absorption bands in different phases, as expected in view of the crudity of the oriented-gas model. One spectral feature which might have been expected to be reasonably well predicted by the oriented-gas model is the direction of the transition moment, since the model stresses orientation of the molecules exclusively. However, even the polarisation ratios calculated using this model are in poor agreement with observed values. (The agreement becomes better, however, if the model is refined by the inclusion of molecular interactions leading to configuration interaction.) Clearly then, the oriented-gas model is useful only as a first approximation.

The simplest way to improve upon this model is to consider a pair of molecules with mutual orientation and separation fixed (e.g. to their values in a crystal of the substance). Spectroscopic changes arising from inter-molecular interaction in this 'oriented physical dimer' model can then be extended to cover the case of an 'infinite' three-dimensional lattice. Suppose the ground-state wave-functions of the two molecules are ψ_1 and ψ_2, so that the ground state (ψ_g) of the oriented physical dimer is approximately

$$\psi_g = \psi_1 \psi_2, \tag{6.2}$$

and let one of the molecules undergo electronic excitation to ψ_1^* or ψ_2^*. In principle, both molecules have equal probability of being excited, so that quantum-mechanically the wave-function of the singly excited system should be written

$$\psi_{e\pm} = (1/\sqrt{2})(\psi_1 \psi_2^* \pm \psi_1^* \psi_2). \tag{6.3}$$

If the two molecules interact, with an interaction energy operator V_{12}, the energies (E_+ and E_-) of the states with wave-functions ψ_{e+} and ψ_{e-} differ by 2β, where β is a resonance interaction energy given by

$$\beta = <\psi_1^* \psi_2 | V_{12} | \psi_1 \psi_2^* >. \tag{6.4}$$

The overall energy-level diagram for the dimer pair is thus as shown in figure 6.3, where the terms W and W' represent the coulombic interaction energies of ground-state and excited-state systems, respectively.

This simple dimer model at once shows that intermolecular interaction leads to both shifts and splittings of absorption bands in the spectrum. These are determined by the values of W and W' (which depend on such factors as polarity and polarisability of the ground- and excited-state species – see chapter 2) and by the resonance energy β (which depends on the overlap of the relevant orbitals of the two molecules). In some cases, one of the two resulting transitions ($+$ or $-$ in figure 6.3) may be forbidden, depending on the orientations of the two molecules. Thus, using short arrows to represent transition dipoles, figure 6.4 shows three possible situations. In the parallel (a) and head-to-tail (b) orientations only the high-energy and low-energy components, respectively, are allowed, while for the oblique orientation (c) both components are allowed. It follows that interactions between two crystallographically inequivalent molecules will lead to a spectroscopic splitting into two allowed transitions, since the transition moments in two inequivalent molecules will not cancel each other. The splitting arising from interaction between crystallographically inequivalent molecules is known as the Davydov splitting and is in the range from a few hundred to several thousand wave numbers for singlet excited states.

6.3.

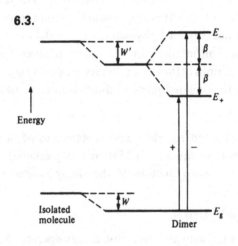

6.4.

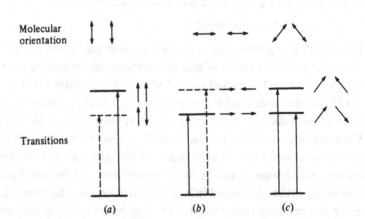

6.3 The exciton concept

Extending the dimer model to an 'infinite' three-dimensional lattice is very similar to the extension of a covalent bond between two atoms to the delocalised band structure in a metal or semiconductor, in that the sharply-defined energy levels of the two-body system become a band of energy levels in the 'infinite' lattice, with the bandwidth dependent on overlap within the lattice. Thus figure 6.3 can be applied to a crystal if each of the levels E_+ and E_- is replaced by a narrow band of levels. These bands are known as crystal exciton bands. For an infinite crystal, equation 6.2 becomes

$$\psi_i' = \psi_i^* \Pi_n \psi_n \tag{6.5}$$

where the *i*th molecule is excited (*) and all other molecules (*n*) are in the ground state. The analogue of equation 6.3, arising from linear combination of the states ψ_i', is then

$$\psi = \sum_{i=1}^{N} a_i \psi_i' \tag{6.6}$$

whose solutions are of the form

$$\psi_k = (1/\sqrt{N}) \sum_{l=1}^{N} e^{ikld} \psi_l' \tag{6.7}$$

where *d* is the intermolecular spacing.

ψ_k is the wave-function of an electronically excited, neutral yet mobile state of the crystal, which is known as an exciton. *k* is the wave-vector of the exciton (c.f. chapter 5) and defines the exciton momentum p ($=\hbar k$). The eigenvalues of the states represented by ψ_k are

$$E(k) = E_0 + (W - W') + 2\beta \cos(kd) \tag{6.8}$$

where β is the energy of interaction between neighbouring molecules and E_0 is the transition energy for the isolated molecule in the gas phase (c.f. figure 6.3).

Since *k* has values 0, $\pm 2\pi/Nd$, $\pm 4\pi/Nd$, ..., $\pm \pi/d$, the bandwidth is equal to 4β.

6.4 Comparison with experiment

Figure 6.5 is an example of how the general predictions of the above models are confirmed in practice[11]. The spectra of three different solid phases of metal-free phthalocyanine are compared with the solution spectrum. The peaks in the solid spectra are shifted and split compared with those in the solution spectrum. Their positions and splittings differ for the different solid phases since the relative orientations of the molecules, and hence the intermolecular interactions, are different in each phase. (Thus electronic spectroscopy can help to identify particular phases of a molecular crystal.) The solid-state peaks are also broader than those in solution since intermolecular interactions depend on the extent of thermal vibrations in the crystal as well as on relative orientation. Cooling a crystal to low temperature reduces vibrational amplitudes, resulting in more sharply defined transitions, and in some cases may also lead to a shift in line position since the splitting of transitions into different components will also depend on temperature via the temperature-dependence of overlap. For example, figure 6.6 shows how the shape and energy of one absorption line of anthracene varies between 20 and 2 K[12]. The limiting linewidth at low temperatures is very sensitive to crystal quality, since defects lead to molecules in a range of relative orientations and hence a broader spectral line than in a perfect crystal. For example, figure 6.7 shows the same line as in figure 6.6 for sublimation and melt-grown crystals of anthracene at 1.8 K. The broader line for the latter crystals reflects their poorer quality.

6.5. Absorption spectra of metal-free phthalocyanine.

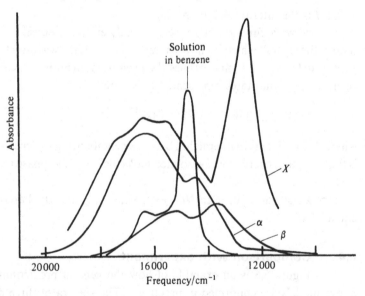

6.5 Coherent and incoherent excitons

Since excitons are generally produced by excitation of electronic transitions of molecular crystals using visible or ultra-violet light of wavelength much longer than the intermolecular spacing, the photon momentum is small and, as discussed in chapter 5 for phonon bands, momentum transfer to the exciton is negligible. Thus, $\Delta k = 0$ and the

6.6. Temperature-dependence of line-shape and line-energy for anthracene crystal.

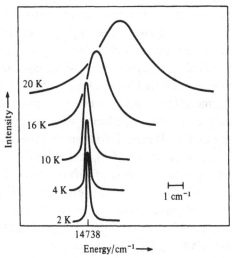

6.7. Comparison of linewidth for sublimation-grown and melt-grown anthracene crystals.

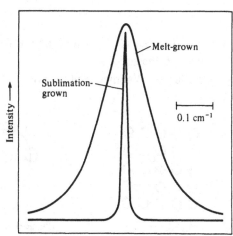

observed absorption bands mainly produce excitons with k near zero. However, excitons may gain or lose momentum by collision with lattice defects or with phonons. The time an exciton stays in a particular k state is known as the coherence time of that state ($\tau(k)$) and the distance it travels during this time is the coherence length, $l(k)$. If $l(k)$ is large compared with the lattice constant, exciton motion is said to be coherent; otherwise it is incoherent. The coherence time is typically $<10^{-14}$ s.

6.6   Types of exciton

In marked contrast to the excited states of an isolated molecule, excitons are not restricted to excitations producing electron–hole pairs located on the same molecular site. Excitons of the latter type are known as Frenkel excitons. In the crystal, the electron may on average be several molecules away from the hole; in this case the resulting exciton is known as a Wannier exciton. An intermediate case occurs when the electron and hole occupy adjacent molecules, a situation producing what is known as a charge-transfer exciton (figure 6.8). Hence, in addition to Frenkel exciton bands with energies described by equation (6.8), molecular crystals have charge-transfer and Wannier exciton bands of different energies. One approach to calculating the energies of Wannier excitons considers the electron as moving in spherical shells of radius r, with hole and electron separated by a medium of effectively uniform dielectric constant. This leads to a series of hydrogen-like Rydberg levels, with energies

$$E = E_\infty - E_1/n^2 \qquad (6.9)$$

where $n = 1, 2, 3$, etc.,

$E_\infty =$ the energy to produce an electron–hole pair with infinite separation ($r = \infty$) in the solid,

and $E_1 =$ the energy of the lowest Wannier exciton level.

6.8.

Frenkel exciton   Wannier exciton   Charge-transfer exciton

For charge-transfer excitons, in which the hole and electron are located on adjacent molecules,

$$E_{CT} = IP - EA - P - C \qquad (6.10)$$

where IP = ionisation potential of donor (hole site),
      EA = electron affinity of acceptor (electron site),
      P = energy of polarisation of lattice by electron–hole pair,
and C = Coulomb attraction energy of electron–hole pair.

These approaches give a good description of observed charge-transfer and Wannier exciton levels considering their relative simplicity. For example, charge-transfer exciton energies in tetracene can be reproduced well using electrostatic coulomb attraction terms for electron–hole pairs with the hole at the origin and the electron centred on neighbouring molecules at (1/2, 1/2, 0); (0, 1, 0); (1, 0, 0); and (1, 1, 0)[13].

In practice, real crystals possess yet more electronic energy levels than those described above, because of the presence of impurities, lattice imperfections and surface states. Impurities with excitation energies lower than the host exciton levels can effectively trap the mobile excitation energy as mentioned earlier, leading to fluorescence characteristic of the impurity rather than of the host. Impurities with higher excitation energy act by disturbing the surrounding host molecules from their normal lattice sites. These and other lattice imperfections (chapter 4) lead to changes in the interaction energies of the perturbed host molecules and their neighbours, which from the above discussion of exciton band energies can readily be seen to produce changes in the excited-state levels in such regions of the crystal. These changes may be in the sense of increasing or decreasing the energy levels relative to those of the perfect host crystal, but only states below the host levels may be readily detected by characteristic fluorescence emission, since states above the host levels will rapidly transfer their excitation energy to lower-lying neighbouring host molecules. Such modified levels are known as X-traps[14]. Molecules at surface sites are not surrounded on all sides by molecules in normal sites and so W, W', β, P, etc. (figure 6.3 and equation 6.10) are all different from those of the bulk molecules, yielding surface states which have been observed directly, for example in naphthalene and anthracene[15].

Electrons and holes in molecular crystals can therefore exist in a whole spectrum of energy states, from the tightly bound electron–hole pairs of Frenkel excitons, through charge-transfer and Wannier excitons to surface states and traps and finally to free mobile charge-carriers. The motion of electrons and holes in molecular crystals can similarly be regarded as pure energy-transfer (in the case of Frenkel exciton motion), motion of tightly or loosely bound ion-pairs (for charge-transfer or Wannier excitons) or pure charge-transport (in the case of motion of trapped or free charge-carriers, which will be discussed in chapter 8).

6.7   Energy transport

The simplest mechanism for energy transport in molecular crystals occurs when fluorescence from one molecule in the crystal is re-absorbed by a molecule on a distant lattice site. Since the excited state of the original molecule and those of other molecules subsequently excited by this process all have a characteristic lifetime, the effect of this emission–re-absorption process is to increase the *apparent* excited-state lifetime in the crystal compared with that in an ideal infinitely thin sample. This increase is also naturally greater for short wavelength components of the fluorescence which are more strongly re-absorbed. Hence this mechanism can be recognised by apparent fluorescence lifetimes which are dependent on crystal thickness and on wavelength. The existence of other energy-transport mechanisms has been demonstrated in several elegant experiments. For example, triplet exciton motion has been studied at 4.2 K in crystals of fully-deuterated benzene containing 0.4% each of C_6H_5D and C_6H_6, whose triplet excited states lie 170 and 200 cm^{-1} below those of C_6D_6[16]. Since the singlet–triplet transition is spin-forbidden, the emission–re-absorption mechanism cannot be invoked and the phosphorescence intensities for C_6H_5D and C_6H_6 would be expected to be equal. In fact an intensity ratio of 1:10 is observed, showing that trap:host, host:host and host:trap energy transfer must have occurred, permitting the expected favoured population of (and therefore emission from) the lower energy trap.

More recently[17], a particularly ingenious method has been used to follow the rate of electronic energy transfer in molecular crystals directly. Two time-coincident Gaussian picosecond pulses of the same wavelength are produced by beam-splitting a pulsed laser source. These are crossed in the bulk of a molecular crystal, at an angle θ, producing a diffraction pattern. Optical absorption then produces a distribution of excited states with exactly the same pattern. A third beam, from the same laser pulse but time-delayed, then probes the crystal. Since the absorption of light by the excited-state molecules to yield higher excited states occurs with different probability from absorption at the same wavelength by the ground-state molecules, the excited-state pattern acts as a diffraction grating for the delayed pulse. However, the grating effectively disappears after a short time, because of the finite lifetime of the excited-state species *and* the transfer of energy from one molecule to new sites. From following the decay of the resulting diffracted-beam intensity by varying the time-delay of the probe pulse, the decay constant can be determined. Measurements as a function of the scattering angle (i.e. fringe spacing) yield the diffusion coefficient and excited-state lifetime directly.

This method has been used to study pentacene in 10^{-3} mol/mol concentration in single crystals of *p*-terphenyl. The decay time of the diffracted intensity (1.5 ns) was much shorter than the value of 4.75 ns predicted if decay solely depended on the lifetime of the excited state, and a diffusion coefficient of 2 cm^2 s^{-1} was found to fit the data well. The method also allows, in principle, experimental distinction between coherent and incoherent motion of excitons, since the decay of the diffraction intensity is exponential for incoherent transport but shows a damped oscillatory non-exponential behaviour in the coherent limit.

The motion of excitons can be considered as lying between two extremes. On the one hand, in the discussion earlier in this chapter the exchange of energy between a pair of strongly-coupled molecules, leading to a splitting in the observed spectrum, was described. Calculations[18] of the rate of exchange, K, assuming coherent transfer, yield

$$K = 4|J|/\hbar \tag{6.11}$$

where J is the strength of the intermolecular interaction (proportional to R^{-3}, where R is the molecular separation). The values of K thus deduced are much higher than observed experimental values, which are commonly of the order of 10^{13} s^{-1}. On the other hand, in the case of weakly-coupled molecules the rate of exchange was deduced by Förster[19] to be proportional to R^{-6}, since the energy transfer is dependent on the product of an inducing field ($\propto R^{-3}$) and the induced dipole moment ($\propto R^{-3}$).

In practice, since the interactions between molecules which are central to the exciton concept are dependent on lattice vibrations and on the extent of any crystal imperfections, excitons maintain coherence for only a very short time ($\sim 10^{-14}$ s) and their motion is best described as intermediate between the extremes above, with diffusive character sometimes regarded as a random walk of a series of hops from one site to another. Consideration of the values of the observed diffusion coefficients together with the likely hopping rates and exciton lifetimes suggests that individual hops may in some cases be longer than nearest-neighbour distances in the crystal lattice[20]. Experimental measurements of diffusion coefficients yield values ranging from 2 cm^2 s^{-1} for picosecond pulses probing into the region verging on coherent transport of singlet excitons[17], down to $10^{-7}\,\mathrm{cm^2\,s^{-1}}$ for triplet excitons in anthracene in incoherent conditions[21]. (These figures are to be compared to $10^{-4}\,\mathrm{cm^2\,s^{-1}}$ for $\mathrm{H^+}$ and $2\times10^{-5}\,\mathrm{cm^2\,s^{-1}}$ for $\mathrm{K^+}$ in water.) Since the diffusion length l (i.e. the rms displacement of the exciton during its lifetime) is related to the diffusion coefficient via $l=\sqrt{(zDt)}$ (where z ranges from 2 to 6 depending on the dimensionality of the motion), excitons can move appreciable distances during their lifetimes (e.g. a singlet exciton with lifetime 10 ns and $D=10^{-4}\,\mathrm{cm^2\,s^{-1}}$ has a diffusion length of about 25 nm, while a triplet exciton of lifetime 10 ms and a similar diffusion coefficient has a diffusion length of 25 μm). This motion may be limited by many processes to which the exciton is subjected during its lifetime: by trapping at impurities or defects; by interaction with the crystal surface; by interaction with another exciton or with a photon or by dissociation of the exciton to yield charge-carriers by processes to be discussed in chapter 8. Exciton motion is also more difficult for charge-transfer and Wannier excitons, which are associated with larger charge separation and hence local polarisation of the surrounding lattice, than for Frenkel excitons. The detailed description of theories of exciton motion[22] is one of the most conceptually and mathematically complex areas of materials science and as such must be strictly avoided in a text of this level. However, the foregoing simple treatment explains a good many of the practical consequences of exciton formation and motion in homomolecular crystals.

6.8   Charge-transfer spectra

Finally in this chapter, the electronic spectra of electron donor–acceptor complexes and charge-transfer salts will be considered. In view of the interest in these materials for their potential molecular electronic applications (to be described in chapter 8), their electronic spectra have been studied more intensively in recent years to obtain information on the electronic structure of the solids, of relevance to semiconducting and metallic properties.

In weak complexes between electron donors and acceptors, the electronic ground state is predominantly neutral with a small degree of charge-transfer character, as predicted by Mulliken's charge-transfer theory[23] outlined in chapter 2. Since the charge transfer is predominantly between neighbouring molecules and localised, the oriented-gas model can be applied quite effectively for the prediction of charge-transfer transition energies, relative intensities and polarisation ratios using, for example, semi-empirical molecular-orbital calculations. If configuration interaction is incorporated in the calculations, the extent of any mixing of charge-transfer with local excitations of the donor or acceptor molecules can be estimated. For example, figure 6.9 shows the polarised spectra of a perylene/tetracyanoethene crystal[24], together with the calculated transition energies and relative intensities[25]. As predicted, the lowest-energy transition, which is largely charge transfer in character, is strongly polarised along the b-axis, the stacking axis in the crystal. The second main absorption feature, in the $20-30000\,\mathrm{cm}^{-1}$ region, consists of several transitions of similar energy, some of which have mixed charge-transfer and local-excitation-of-donor character. Since local excitations of perylene are polarised predominantly in the molecular plane whereas charge-transfer transitions are polarised along the stacking axis, these mixed-character transitions are less strongly polarised than the lowest-energy transition. Similar measurements have been made for many other charge-transfer complexes.

As discussed in chapters 2 and 3, charge-transfer interactions depend on

6.9. Polarised crystal spectra (with theoretical predictions, dotted lines) for perylene/tetracyanoethene.

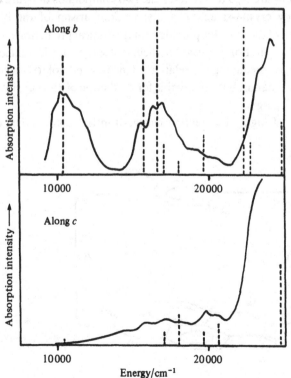

orbital overlap between the electron donor and acceptor and as such have conformational requirements. The study of polarised crystal spectra of charge-transfer complexes provides a means of experimentally testing theories of these conformational requirements. Since crystalline charge-transfer complexes between particular donors and acceptors generally form only one stable phase at room temperature, if two crystalline compounds are to be obtained in which the same donor and acceptor are constrained in different relative orientations it is necessary to force the orientational constraint by chemical means.

One approach to this has involved the use of polymethylene chains of various lengths to link donor and acceptor moieties in compounds known as cyclophanes (e.g. figure 6.10). The polymethylene chains must be long enough to avoid any ring strain which might distort the planarity of the donor and acceptor moieties, and the conformational flexibility of such long chains means that in solution the molecules do not have a single fixed donor–acceptor orientation. In crystals, however, both intra- and inter-molecular charge-transfer interactions occur, involving different relative orientations and separations of donor and acceptor. These orientations and separations are also different for the two compounds of figure 6.10, as determined by crystal-structure analysis. Comparison of the observed crystal spectra of the two compounds with theoretical predictions based on the observed orientations shows[26] that current semi-empirical calculations are adequate for predicting the relative intensities and polarisations of the charge-transfer bands, but not reliable for prediction of absolute intensities or transition energies.

In crystals of highly conducting charge-transfer salts, studies of elec-

6.10.

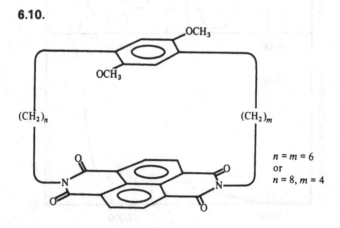

$n = m = 6$
or
$n = 8, m = 4$

tronic spectra are extremely valuable both for distinguishing semiconducting and metallic character and for probing the dimensionality of the charge transfer. Thus in many TCNQ salts, low-energy absorption bands strongly polarised along the TCNQ stacking axis are observed[27] (e.g. figure 6.11), corresponding to charge transfer between TCNQ molecules in various stages of negative charge. This is in marked contrast to the metallic reflectivity observed, for example, in tetramethyltetraselenafulvalene hexafluorophosphate (($TMTSF)_2PF_6$) (figure 6.12). In the latter case, a sharp rise in reflectivity is observed at a particular frequency, which may be explained in terms of the Drude theory of metals and from which a value for the optical effective mass and hence the conduction bandwidth may be derived[28]. This reflectivity is anisotropic and temperature-dependent, with the sharp rise being more pronounced at lower temperatures and for light polarised along the highly conducting stacking axis *a* than for light polarised perpendicular to this direction along *b*. These results are

6.11. Cs_2TCNQ_3 crystal structure and polarised crystal spectra.

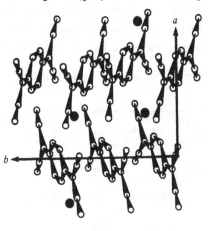

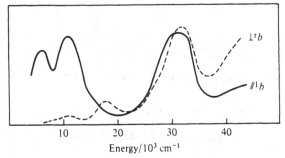

Energy/10^3 cm^{-1}

consistent with one-dimensional metallic character with a trend towards two-dimensional character at low temperatures. Comparisons of the anisotropy of metallic reflectivity for different materials of this class thus provide valuable information on the role of inter-stack contacts in promoting a degree of two-dimensional character in the band structure. In addition to re-inforcing electrical conductivity data, these measurements are simpler experimentally than the measurement of conduction aniso-tropy. The electrical properties of these materials are discussed in chapter 8.

6.9   High-technology applications

Recently the optical properties of organic solids have found a wide range of high-technology applications[29,30]. These range from infra-red absorbers for use in 'invisible' bar-codes for security applications and in write-once-read-many-times (WORM) memory systems, through colour microfilters for liquid crystal colour displays, to flash-protection filters and other non-linear optic materials, and materials for electrophotography and solar photovoltaic cells. Non-linear optic materials are covered in chapter 10, while electrophotography and solar cell applications are discussed in section 8.9. The other applications mentioned above require identification

6.12. $(TMTSF)_2PF_6$ crystal structure and polarised reflectance spectra.

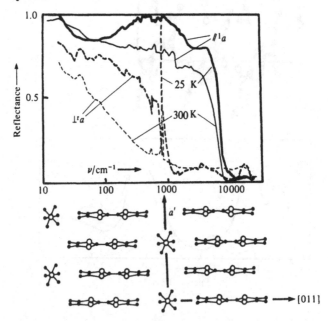

of molecules having the correct spectroscopic properties (which can be done using conventional molecular orbital calculations); synthesis of candidate molecules and development of methods for fabrication of films or other suitable forms of the colorant; determination of solid-state spectroscopic properties; and a range of associated property evaluations including stability and toxicity testing.

WORM memory systems provide very high-density information storage using the same principle as the compact disc except that the optical features which store the information to be read by a low-power diode laser are produced by high-power-laser-ablation of a dye layer on a reflecting metal substrate. Phthalocyanines are commonly used as the dye layer on account of their excellent thermal and light stability. However it is necessary to shift their normal absorption (see figure 6.5) to a region centred on 820 nm ($12\,200\,\text{cm}^{-1}$) and to render the material soluble for solvent- or spin-coating. Both these requirements can be achieved by near-complete peripheral substitution of the phthalocyanine with thioaryl groups. This is because of the electronic effect of these substituents on the π-electron system, together with the steric hindrance which forces them out of the phthalocyanine ring plane (with consequent reduction in interaction of π-systems of adjacent molecules and hence sharper spectral bands and increased solubility). Thin films of such dyes on metal-coated melinex substrates and coated with a protective overlayer of polyacrylate are used in a commercial system combining high writing speed, good long-term stability and cheap production costs[31].

Phthalocyanines again find application in colour microfilter screens for colour liquid crystal displays. Such filters must provide clearly defined red, green and blue arrays of sub-pixel dimensions. One method of fabrication consists of producing a mask pattern in photoresist, which is then sublimation-coated with an insoluble organic pigment. Solvent dissolution of the photoresist lifts off the dye in all regions except those where the original photoresist had been etched away prior to coating with the pigment. Repetition of the process with the remaining pigments builds up the necessary filter array. Copper phthalocyanine (blue), octaphenyl copper phthalocyanine (green) and peryleneimide dyes (red) have proved suitable.

Some organic dyes show optical extinction coefficients which rise very rapidly at high light intensities and can be used as optical switches or flash protection devices. One phenomenon which leads to such effects is very strong optical absorption from a long-lived first excited state. At low light intensities, the first excited state is populated relatively slowly, so the equilibrium excited state concentration is low and absorption follows the Beer–Lambert law. However, at high light intensities the population of the first excited state becomes very high, leading to onset of the new intense absorption to higher states. This is known as reverse saturable absorption and is one example of non-linear optical properties, i.e. optical properties which vary in a non-linear manner as light intensity is increased. These properties are discussed in more detail in chapter 10.

All of these optical applications of organic solids involve materials characterisation which demands understanding of the methods and theoretical concepts developed earlier in this chapter.

References

1 G.F.A. Körtum, *Reflectance Spectroscopy; Principles, Methods, Applications*, Berlin: Springer, 1969.
2 K. Yakushi, M. Iguchi and H. Kuroda, *Bull. Chem. Soc. Japan*, 1979, **52**, 3180.
3 R.K. Ahrenkiel, *J. Opt. Soc. Am.*, 1971, **61**, 1651.
4 A. Rosencwaig, *Photoacoustics and Photoacoustic Spectroscopy*, New York: Wiley, 1980.
5 G.F. Kirkbright and S.L. Castleden, *Chemistry in Britain*, 1980, **16**, 661.
6 H.H. Perkampus, *Naturwiss.*, 1982, **69**, 162.
7 N.J. Bridge and L.P. Gianneschi, *J. Chem. Soc. Faraday II*, 1976, **72**, 1622.
8 A. Winterstein, U. Schön and H. Vetter, *Naturwiss.*, 1934, **22**, 237.
9 P. Pringsheim and A. Kronenberger, *Z. Physik.*, 1926, **40**, 75.
10 H.C. Wolf, *Z. Naturforsch.*, 1959, **A13**, 414.
11 J.H. Sharp and M. Landon, *J. Phys. Chem.*, 1968, **72**, 3230.
12 H. Port, K. Mistelberger and D. Rund, *Mol. Cryst. Liq. Cryst.*, 1979, **50**, 11.
13 L. Sebastian, G. Weiser and H. Bässler, *Chem. Phys.*, 1981, **61**, 125.
14 J.O. Williams and J.M. Thomas, *Surface and Defect Properties of Solids Specialist Periodical Reports*, London: Chemical Society, 1973, **2**, 229.
15 M.R. Philpott and J.M. Turlet, *J. Chem. Phys.*, 1976, **64**, 3852.
16 G.C. Nieman and G.W. Robinson, *J. Chem. Phys.*, 1963, **39**, 1298.
17 J.R. Salcedo, A.E. Siegman, D.D. Dlott and M.D. Fayer, *Phys. Rev. Lett.*, 1978, **41**, 131.
18 J. Perrin, *C. R. Acad. Sci. (Paris)*, 1927, **184**, 1097. F. Perrin, *Ann. Chim. Physique*, 1932, **17**, 283.
19 T. Förster, in *Modern Quantum Chemistry, Part 2: Action of Light on Organic Molecules*, ed. O. Sinanoglu, New York: Academic Press, 1965, p. 93.
20 N.J. Bridge and D.P. Solomons, *J. Chem. Soc. Faraday II*, 1980, **76**, 472.
21 B. Nickel and H. Maxdorf, *Chem. Phys. Lett.*, 1971, **9**, 555.
22 See, for example, V.M. Kenkre and R.S. Knox, *Phys. Rev.*, 1974, **B9**, 5279.
23 R.S. Mulliken, *J. Am. Chem. Soc.*, 1952, **74**, 811.
24 H. Kuroda, T. Kunii, S. Hiroma and H. Akamatu, *J. Mol. Spectroscopy*, 1967, **22**, 60.
25 T. Ohta, H. Kuroda and T.L. Kunii, *Theoret. Chim. Acta (Berlin)*, 1970, **19**, 167.
26 R.L.J. Zsom, L.G. Schroff, C.J. Bakker, J.W. Verhoeven, Th.J. de Boer, J.D. Wright and H. Kuroda, *Tetrahedron*, 1978, **34**, 3225.
27 J. Tanaka, M. Tanaka, T. Kawaii, T. Takabe and O. Maki, *Bull. Chem. Soc. Japan*, 1976, **49**, 2358.

28 C.S. Jacobsen, D.B. Tanner and K. Bechgaard, *Phys. Rev. Lett.*, 1981, **46**, 1142.
29 P. Gregory, *High-technology Applications of Organic Colorants*, New York: Plenum, 1991.
30 H. Böttcher, T. Fritz and J.D. Wright, *J. Mater. Chem.*, 1993, **3**, 1187.
31 See *New Scientist*, 25 February 1988, p. 37; *Chemistry and Industry*, 7 March 1988, p. 133.

7

Chemical reactions in molecular crystals

Chemical reactions between molecules in crystals differ in many respects from reactions between the same molecules in solution or gas phases. Most obviously, in the crystalline state the molecules possess little or no kinetic energy, so the majority of organic solid-state reactions are photochemical rather than thermal. Solid-state reactions are frequently far more stereo-specific and free from side-reactions (e.g. with solvent) than are reactions in solution. This is because many reactions of molecular crystals take place under topochemical control, i.e. in conditions where the intrinsic reactivity of the molecule is less important than the nature of the packing of the neighbouring molecules around it in the crystal. In these conditions, the separation, orientation and permissible approach-directions of reactive regions of adjacent molecules are well-defined and controlled by the lattice structure of the material, so that single, specific reaction products are predetermined. However, the range of organic solid-state reactions and the degree of control which can be exercised using solid-state conditions is far wider and far more subtle than would be the case if each substance crystallised in a single structure with molecules rigidly fixed in a perfect lattice.

As has been shown in chapter 3, the weak intermolecular forces in molecular crystals frequently permit the formation of several phases with different crystal packing. Each of these polymorphs will in general involve a unique relative orientation and separation of adjacent molecules, permitting control of reaction pathways and products by selection of the appropriate phase. Also, the presence of defects (chapter 4) and the occurrence of molecular motion (chapter 5) further modify the range of possible reaction pathways so that the products are sometimes not those predicted by the perfect static lattice.

Before these features are considered in detail, it is important to note that not all organic solid-state reactions occur under topochemical control. There are many reactions of both inorganic and organic solids which occur by 'reconstructive' mechanisms, in which reaction starts at a few nucleation centres, grows outwards from these at an increasing rate as the area of the reaction front increases and finally slows down as the reaction fronts from adjacent nucleation centres merge.

7.1 Non-topochemical reactions in molecular crystals

Photochemical polymerisations of vinyl monomers are good examples of reactions which do not occur under topochemical control[1]. In such reactions, large geometric changes occur, since two trigonally hybridised carbon atoms are converted to a tetrahedrally hybridised state (figure 7.1). X-ray diffraction studies have shown that even at only 4% polymerisation of monomer crystals, the polymer is amorphous and forms a separate phase within the monomer crystal[2]. Photomicrographs of the surfaces of acrylamide ($X = CONH_2$) crystals during polymerisation show that polymer globules develop, initially about 20 Å diameter, along lines oriented in particular crystallographic directions and thought to be dislocation lines. This suggests that excitons formed by the initial absorption of ultra-violet light migrate through the crystal until they become trapped at dislocations, where they lead to radical formation and polymer initiation. (The alternative of direct excitation of molecules in the vicinity of dislocations can be shown to play a negligible role since the rate of initiation of the reaction is strongly dependent on the direction of polarisation of the incident light with respect to the crystal axes, as expected on the exciton model with excitons originating at molecules in normal ordered lattice sites, but not expected for direct excitation of less-well-ordered molecules near dislocations.) The amorphous globules grow to a diameter of 300–400 Å, during which time the structural change associated with the polymerisation sets up strains in the crystal, producing new dislocations at which further nucleation of polymerisation, following exciton trapping, can occur. In addition to acting as an exciton trapping site, the region of lattice disruption surrounding a dislocation provides

7.1.

maximum facility for accommodating the large structural change occurring on polymerisation. If illumination is cut off during the polymerisation, it is found that the reaction continues at a slower rate, with approximately constant free-radical concentration. Under these conditions, migration of existing radicals to new dislocations occurs slowly, followed by a temporary acceleration and repetition of the cycle.

Although reactions of this type clearly show the importance of the exciton concept and the role of structural defects in solid-state reactions, their significance for controlled polymer synthesis is limited. The product is an amorphous polymer whose structure is not related to that of the original monomer crystal, and since polymerisation originates at defect sites it is difficult to modify in any controlled way. However, it is possible to impose a degree of crystal-lattice control on reactions of this general type if the molecules of the monomer are trapped in channels in the clathrate structures of urea or thiourea. Thus, for example, γ-ray-initiated polymerisations of such clathrates of 2,3-dimethyl- and 2,3-dichloro-1,3-butadiene and of 1,3-cyclohexadiene produce all-*trans*-1,4 polymers as needles with the polymer chains aligned along the needle axis[3]. Although the scope of such reactions is limited by the solubility of the molecule forming the host clathrate, they may be regarded as organic analogues of the zeolite systems in some respects, with the monomer molecules constrained enough to give controlled product structures but not so rigidly as to restrict the structural change inevitably occurring on polymerisation. They are, therefore, on the borderline of topochemical reactions. True topochemical reactions involve reactions of molecules whose structures and/or relative orientations are controlled by crystal lattice structure.

7.2   Influence of crystal-controlled molecular structure on chemical reactivity

There are relatively few clear-cut examples of reactions where lattice control of just the molecular structure is involved, since intermolecular interactions nearly always play a part directly as well as indirectly via their influence on molecular conformation. One example is the reaction of crystals of *meso*-2,3-dibromo-1,4-butanedicarboxylic acid dimethyl ester with amines to yield *trans-trans*-1,3-butadiene 1,4-dicarboxylic acid dimethyl ester via a stereospecific *trans*-elimination of HBr[4] (figure 7.2).

In the solid state, the fixed molecular conformation leads to a single product, whereas the corresponding reaction done in solution yields a mixture of the *trans-trans*, *trans-cis* and *cis-cis* isomers as well as some amide by-products. These products of solution reaction also reflect *trans*-

elimination of HBr, but in this case from a range of conformations of the flexible molecule in solution rather than from a single fixed conformation.

The effects of lattice restriction of intramolecular motion may be more subtle in cases where the conformation of an intermediate, rather than of the initial reactant, is controlled by the lattice. For example, the photo-elimination of carbon monoxide from a series of *cis-* and *trans-*1,3-diphenyl 2-indanones occurs via a biradical mechanism[4] (figure 7.3).

In the solid state, internal rotation and inversion at the radical centres is restricted and the product largely retains the stereochemistry of the reactant. However, in solution such rotation and inversion is facile, leading to the least sterically-hindered product, which is the *trans* isomer. If $R = CH_3$ this rotation is hindered, so some degree of retention of stereochemistry is observed even in solution (see table 7.1). Interestingly, in this case the effects of local environment in restricting inversion and rotation are already visible when the reaction is carried out in viscous

7.2.

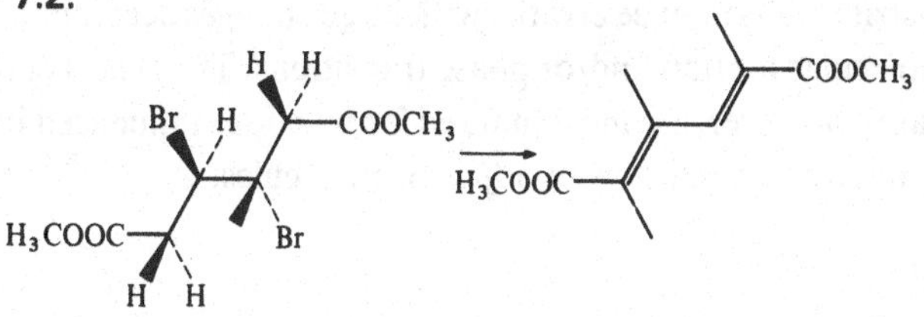

Table 7.1 *Photoelimination reactions of 1,3-diphenyl-2-indanones*

Reactant	Solid products	Solution products
R = H, *cis*	95% *cis*	11% *cis*
	5% *trans*	89% *trans*
R = H, *trans*	5% *cis*	13% *cis*
	95% *trans*	87% *trans*
R = CH$_3$, *cis*	90% *cis*	69% *cis*
	4% *trans**	31% *trans*
R = CH$_3$, *trans*	14% *cis*	9% *cis*
	86% *trans*	91% *trans*

*6% of 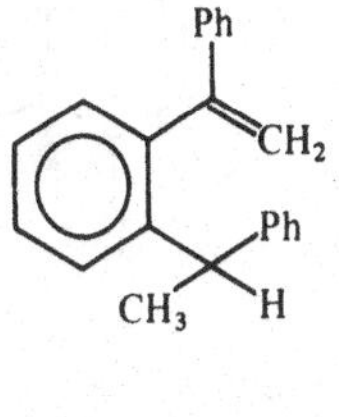   also formed

solvents, which give product ratios intermediate between those of solution and solid reactions.

In extreme cases, solid-state influences on molecular reactivity may be so dramatic that reaction is observed only in the solid state and not in solution. For example, the epoxy alcohol *A* is stable in solution but its crystals liquefy on standing, yielding a mixture of *B* and *C* (figure 7.4)[6]. In this case the solid-state effects are not entirely steric, since the crystals react much more slowly when stood in ether. (It has been suggested that the reaction is acid-catalysed, since the solution reaction will occur in ether in the presence of HBr and ethanoic acid, and that the ether stabilises the crystals by removing HBr from the reaction site, preventing the catalysis.)

Two general characteristics of all solid-state reactions involving lattice control of molecular conformation concern substituent effects and temperature-dependence. Substituent effects in such reactions are expected to correlate with steric rather than electronic characteristics of the substituents. Finally, the temperature-dependence of reaction rate and product distributions may be erratic, with large changes accompanying the onset of molecular motion and/or phase transitions. The effects of different crystal phases, however, are most marked for reactions influenced by lattice control of relative orientation of adjacent molecules.

7.3.

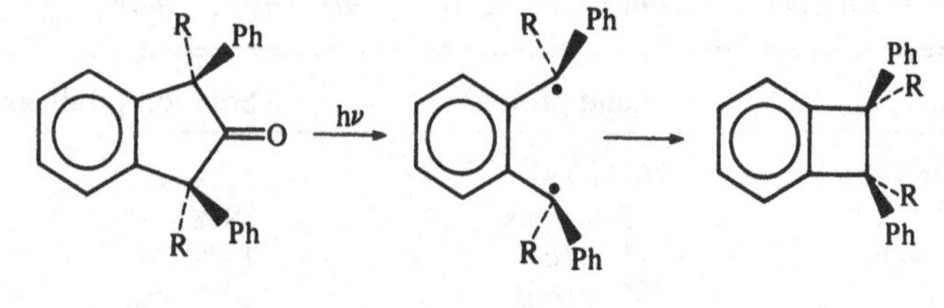

7.4.

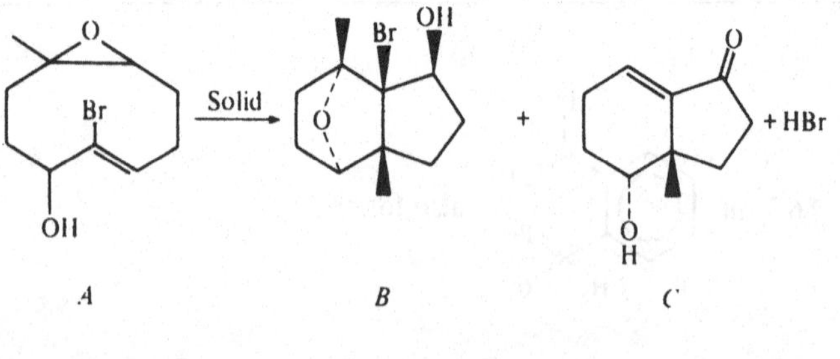

7.3   Influence of crystal-controlled relative orientation of molecules on chemical reactivity

As long ago as 1889 it was observed[7] that irradiation of crystals of *trans*-cinnamic acid (3-phenyl propenoic acid) led to the formation of dimers, whereas the solution or molten states at most isomerised, given the same exposure. At the time it was impossible to rationalise such phenomena, for the technique of X-ray crystallography was still unknown, so no information was available on the molecular shape or environment in the crystal. Later[8] it was realised that the related 3(2-ethoxyphenyl) propenoic acid crystallised in three polymorphic forms: the α-phase, from acetone, yielding a centrosymmetric dimer; the β-phase, from benzene, yielding a dimer having mirror symmetry; and the γ-phase, from aqueous ethanol, which is stable to light. Subsequent determination of the crystal structures of these three phases yielded the results shown in figure 7.5.

Examination of these structures shows that the α- and β-phases place reactive double bonds parallel and approximately 4 Å apart, in a configur-

7.5.

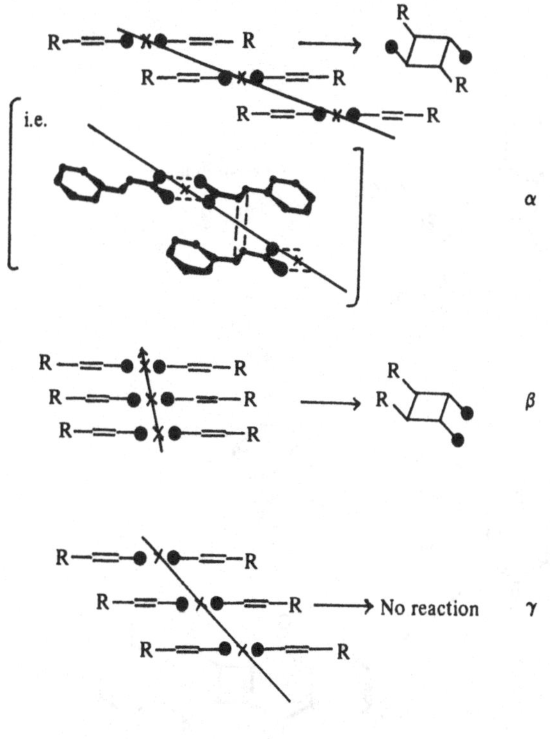

(● = COOH)

ation involving minimal structural change to achieve dimerisation to give the observed products. The γ form, in which the double-bond separation is around 5 Å with large transverse displacement of adjacent double bonds, cannot react because extremely large structural change would be involved. If the double bonds are not parallel, reaction is also forbidden since large structural change would again be required to form a stable dimer. Thus, 3(2-bromophenyl) propenoic acid methyl ester crystallises with the adjacent double bonds only 3.93 Å apart but far from parallel (figure 7.6) and gives no photodimer[9]. (An alternative view of the origin of the lack of reactivity for systems with large transverse displacements and/or non-parallel double bonds is that orbital overlap and symmetry considerations are unfavourable in these cases.)

The requirement of parallel double bonds of the order of 4 Å apart in order for photodimerisation to occur is quite general. For example, $5\alpha,8\alpha$-dimethyl-tetrahydronaphthoquinone undergoes intramolecular cyclisation in solution, but in the solid state, where the double bonds of adjacent molecules are parallel and 3.624 Å apart across a centre of symmetry, a stereospecific dimerisation occurs[10] (figure 7.7).

7.6.

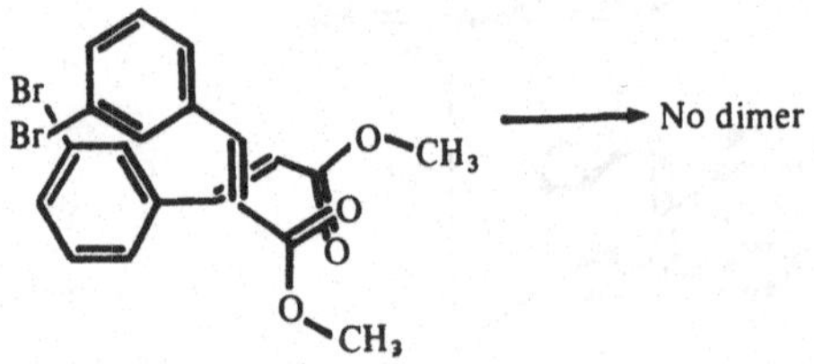

7.7.

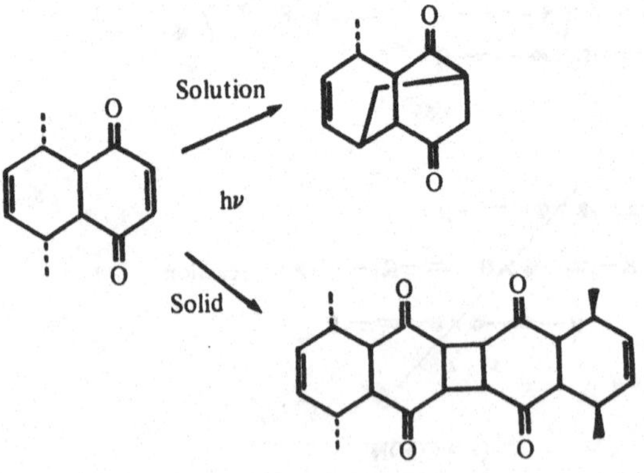

In molecules containing two or more reactive double bonds, solid-state reactions of the above type can lead to stereospecific polymerisation. Thus, distyrylpyrazine (α-phase) crystals from solution (though not those of the γ-phase from vapour growth) photopolymerise even at $-60\,°C$, yielding polymer chains oriented along the c-direction[11]. The product of this solid-state polymerisation has a molecular weight typically $100\times$ larger than that of solution-polymerised material, since the polymer chains propagate smoothly in one dimension with minimal opportunity for chain-termination reactions. Furthermore, the polymer is crystalline and there is a close correspondence between the unit-cell dimensions of the monomer and polymer[12], as expected in a topochemical process involving minimum structural change between reactant and product (figure 7.8 and table 7.2).

The temperature-dependence of these polymerisation reactions reflects their topochemical nature. Thus, 1,4-phenylenediacrylic acid diethyl ester (PDAEt) polymerises quantitatively with nearly zero activation energy

Table 7.2 *Comparison of unit cells of monomer and polymer for some solid state photopolymerisations*

Compound	Space group	a (α)	b (β)	c(/Å) (γ)(/°)	Inter-double-bond distance/Å
2,5-Distyrylpyrazine (DSP) (α-phase)					
Monomer	Pbca	20.638	9.599	7.655	3.939
Polymer		18.36	10.88	7.52	
1,4-Phenylenediacrylic acid dimethyl ester (PDAMe)					
Monomer	P$\bar{1}$	7.148 (98.97)	8.382 (116.85)	5.844 (78.06)	3.957
Polymer	P$\bar{1}$	7.82 (107.8)	7.42 (106.0)	6.04 (78.8)	
1,4-Phenylenediacrylic acid diphenyl ester (PDAPh)					
Monomer	P2$_1$/c	6.917	18.584 (101.87)	7.557	3.917
Polymer	P2$_1$/c	7.50	17.3 (102.0)	7.50	

DSP: ⟨phenyl⟩—CH=CH—⟨pyrazine(N)⟩—CH=CH—⟨phenyl⟩

PDAMe: MeOOC—CH=CH—⟨phenyl⟩—CH=CH—COOMe
(Ph)   (Ph)   (Ph)

between 4.2 and 90 K whereas between 100–170 K the activation energy is $6.7 \pm 1.3 \, \text{kJ mol}^{-1}$, and at higher temperatures, where the lattice is no longer sufficiently rigid to maintain the required action geometry, the yield decreases[13]. The zero activation energy at low temperatures shows that the reaction is a true photochemical process involving monomers reacting to give polymers with minimal structural change. Another consequence of the similarity of monomer and polymer crystal structures is that the crystalline polymers do not melt on heating but undergo thermal depolymerisation yielding the original crystalline monomer.

A related reaction is the solid-state polymerisation of diacetylenes[14], which may be initiated in a variety of ways: by radiation (ultra-violet, X-rays, γ-rays); thermally; chemically; or even by shear deformation of the monomer crystal. The reaction takes place uniformly even in regions of the monomer crystals shown by topography to be free of defects, suggesting a true topochemical mechanism[15]. Application of the empirical criterion proposed by Schmidt[16], that reacting carbon atoms on adjacent monomers should be separated by less than 4 Å if reaction is to be possible, results in limits on the range of lattice geometries for which polymerisation is allowed. These limits are generally expressed in terms of the parameters d and γ in figure 7.9. Figure 7.10 shows the calculated limits of d and γ above

7.8.

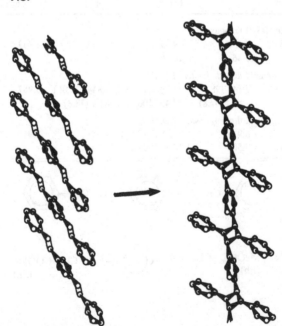

which the separation S is greater than 4 Å. A second, lower, limit is imposed by the closest-approach distance of monomers determined by intermolecular repulsion at around 3.4 Å. There is a large and growing amount of structural and solid-state reactivity data on substituted diacetylenes, some of which is presented in figure 7.10, showing that the observed reactivity limits correspond well with the theoretical predictions[15,17].

7.9.

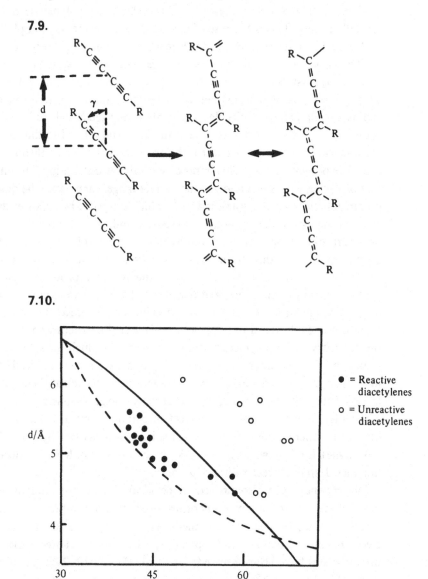

7.10.

● = Reactive diacetylenes

○ = Unreactive diacetylenes

These studies are complicated in practice by the fact that many diacetylene monomer crystals tend to polymerise on exposure to X-rays, so that it is not always possible to determine the crystal structure of the monomer. Low-temperature X-ray diffraction studies, or the use of neutron diffraction, help to minimise such radiation-induced polymerisation during collection of structural data. Despite the extent of these studies, it remains difficult to predict whether a given substituted diacetylene will polymerise in the crystalline state. This is because of the existence of several crystal phases, each with different structure and hence reactivity, for many of the materials.

The reason for the strong interest in these materials is that in favourable cases single crystals of the monomer can be observed transforming directly to single crystals of the polymer which contain very long well-ordered chains of conjugated molecules with interesting electrical and optical properties[18]. If the polymer is to be a single crystal related to that of the monomer, the lattice must control the reaction at all stages through to complete polymerisation. This imposes several new criteria in addition to that of having values of d and γ in the correct range. First, since the repeat unit in polydiacetylene crystals is 4.91 Å, which is significantly shorter than that for many of the monomer crystals, significant strain occurs on polymerisation. Unless the interactions between stacks of monomers in the crystals are strong, this strain remains localised and causes cleavage and hence fibrous products. Second, clearly the polymer product must be topotactic, with orientation fixed relative to that of the monomers in the original parent lattice. Finally, the reaction must not proceed at such a rate that the heat evolved causes local melting of the lattice structure controlling the geometry of the polymerisation. Experiments in which the substituent groups on the monomer diacetylene are varied have helped to clarify the molecular structure requirements for these criteria to be satisfied. Thus, large substituents with strong dipolar character (e.g. toluene sulphonate, carbazole) favour not only the correct orientations but also the correct inter-stack interactions, whereas small substituents (as in 1-hydroxy-2,4-hexadiyne) lead to weak inter-stack interactions and hence a fibrous polymer despite correct values of d and γ.

Diacetylenes have also proved to be ideal systems for studying the detailed mechanism of low-temperature solid-state photo-polymerisation reactions, since the reactive intermediates may be well-characterised by electron spin resonance and optical spectroscopy[19]. Three series of intermediates have been observed: diradicals (DR_n) with $2 \geq n \geq 5$; dicarbenes (DC_n) with $n \geq 8$; and asymmetric carbenes (AC_n); in addition to stable oligomers (SO_n). The electronic structures of each of these species are distinct and, since all involve delocalised π-electrons, they depend on chain length in a way which can be successfully modelled using a one-dimensional electron gas ('particle in a box') approach. This gives the energies (E_n) of the optical absorptions as a function of conjugation length, l, as follows:

$$E_n = (h^2/8ml^2)(4n+1) + E_\infty[1-(1/4n)] \tag{7.1}$$

where E_∞ is the value for an infinite chain, $h =$ Planck's constant, $m =$ electron mass and $l = a_1 n + a_2$ (with $a_1 =$ the length of the repeat unit $\{(2d_{C-C} + d_{C\equiv C} + d_{C=C})$ for SO_n and AC_n, $(d_{C-C} + 3d_{C=C})$ for $DR_n\}$ and a_2 a variable boundary-condition constant). As figure 7.11 shows, the theory is in good agreement with experiment despite its simplicity (only two adjustable parameters, a_2 and E_∞, are involved).

The initial step of the reaction involves formation of an excited diradical species, characterised by ESR, from the singlet excited state of the monomer. This reacts with vibrationally displaced neighbouring molecules to give dimer, trimer, etc. diradicals (figure 7.12) which are also detected by their ESR and optical absorption. Irradiation at the wavelength of absorption of a single member of the diradical series leads to production of the asymmetric carbene and stable oligomer members with additional monomers added to the chain

$$DR_2 + M \xrightarrow{h\nu} AC_{n+1}$$

$$AC_{n+1} + M \underset{h\nu}{\overset{h\nu}{\longrightarrow}} AC_{n+2}$$
$$SO_{n+2}$$

Low temperatures must be used to slow down the reaction enough for observation of the early oligomers. Thus, for studies of dimer species, liquid-helium temperatures are used, while at 100 K and 300 K dimers typically form trimers in times of the order of 1 hour and 1 μs, respectively.

7.11.

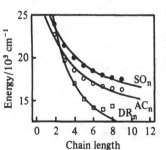

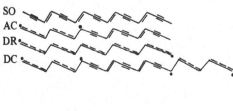

The transformation of diradicals to dicarbenes for chain lengths above six has been explained in terms of electronic structure as follows. Starting from the dicarbene form, a p_z-electron is allowed to become delocalised by successive movement along the chain. Figure 7.13 compares this process for polydiacetylene and polyacetylene. As can be seen, the p_z-electron represents a phase boundary between butatriene and acetylene regions in the polydiacetylene. These regions, in contrast to the situation for polyacetylene, are not of the same energy, and an energy ε is needed to move the p_z-electron along the chain by two carbon atoms as the chain length of the high-energy butatriene segment increases at the expense of the acetylene segment. When the electron approaches the other end of the chain it eventually encounters the other carbene unit of the dicarbene, and recombination of the two p_z-electrons occurs forming an additional π-bond with energy ε_π (figure 7.14). If the energy gained on formation of this new π-bond is sufficient to compensate for the energy required to move the electron, forming more of the high-energy buta-triene segment in the process, then diradicals will be favoured. This is possible for chain lengths $n \leq 5$. For longer chain lengths, the larger distance of movement leads to too unfavourable an energy change, and the energy of formation of a long segment of butatriene structure can no longer be compensated by the formation energy of the new

7.12.

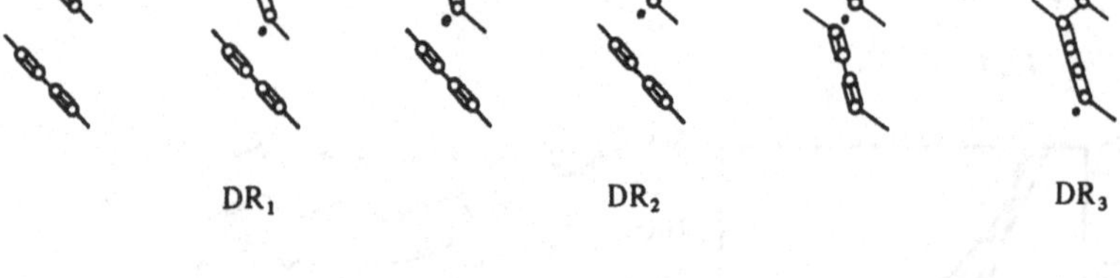

(A = electronic excitation, B = vibrational excitation, C = addition.)

7.13.

double bond. The dicarbene structure is thus stabilised for long chains.

The chain termination reaction is also believed to involve these dicarbene species. In the system studied in most detail, the diacetylene has toluene sulphonate groups as substituents. Intramolecular proton transfer reactions lead successively to asymmetric carbene and stable oligomer products (figure 7.15). These chain-termination steps are consistent with the observation that deuteration of the relevent CH_2 groups slows down the termination reaction. Hence, the entire course of this solid-state polymerisation has been mapped out.

7.4   Controlled topochemical reactions

The elegance and specificity of topochemical reactions such as those described above inevitably raise the question of whether it is possible to utilise known structural principles such as those in chapter 3, together

7.14.

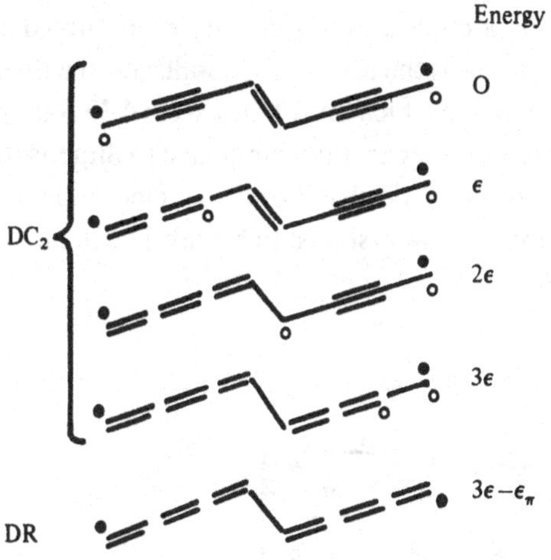

7.15.

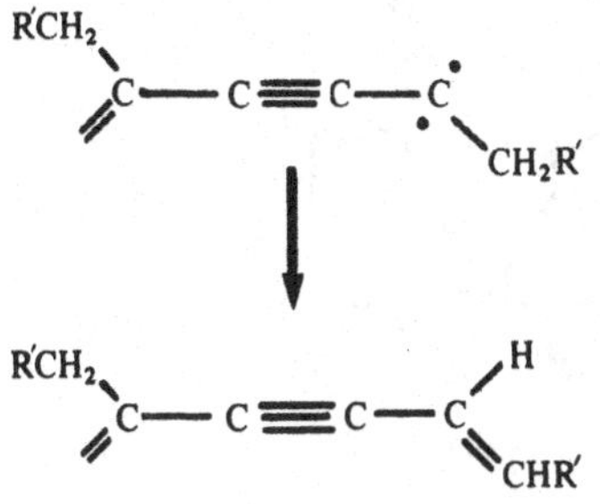

with appropriate choice of molecular structure (e.g. via substituents), to deliberately create solid structures favouring particular topochemical reactions. Although the common occurrence of polymorphism is a severe obstacle to this 'crystal engineering'[8], the increasing understanding of the principles governing structures of molecular crystals, based on vast structural data together with theoretical models capable of predicting those lattice structures having minimum energy, suggest that this approach may be fruitful. This is particularly likely to be the case in structures composed of molecules which interact by strongly orientation-dependent forces, notably dipolar interactions (particularly hydrogen bonding) and charge-transfer interactions.

The simplest and crudest approach to crystal engineering involves the reacting molecules being doped into the lattice of, or co-crystallised with, a substance which dominates the structure of the mixed material and which constrains the reacting molecules approximately in the required orientation for reaction. The example of polymerisation of dienes trapped in the channels of urea or thiourea clathrates has already been quoted earlier in this chapter. A further example is the use of co-crystallisation with mercuric chloride, which has a unit-cell in which one of the axes is 4.33 Å long. This is comparable to the double-bond separation conducive to dimerisation, and several molecules with reactive double bonds in fact undergo totally stereospecific dimerisation on co-crystallisation with $HgCl_2$, for example coumarin[20] (figure 7.16).

7.16.

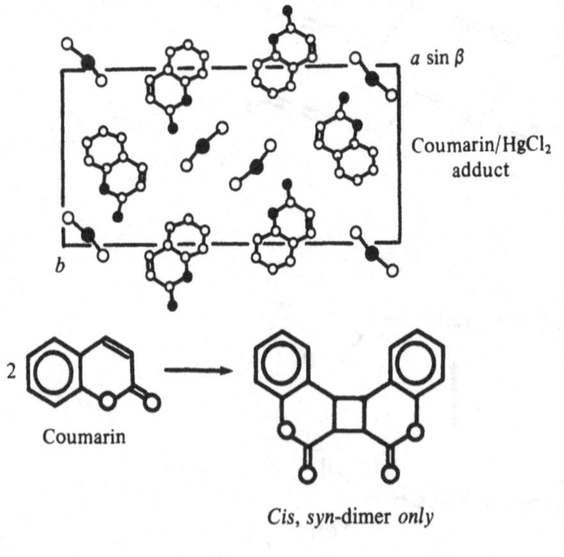

a sin *β*

Coumarin/HgCl₂ adduct

2   Coumarin

Cis, syn-dimer *only*

Another equally simple yet effective approach is known as the 'dichloro rule'[21]. This states that aromatic and related molecules which are capable of crystallising in closely-overlapped head-to-head stacked structures (e.g. of the β-cinnamic acid type, figure 7.5) with the shortest unit-cell axis having a length 4 ± 0.2 Å, will invariably do so if the molecule contains two chloro substituents. This principle is valid for a wide range of molecules, with the two chlorine atoms in various positions, although dichlorophenyl is a particularly useful functional group to exploit the effect. Thus, for example, co-crystallisation of the two substituted butadienes in figure 7.17 yields a crystal which undergoes stereospecific photodimerisation to produce the optically-active product shown. In this reaction the formation of symmetrical dimers of the phenyl compound is avoided by selective excitation at the longer wavelengths absorbed only by the thienyl compound, while the use of low concentrations of the latter ensures that few thienyl symmetric dimers are formed. One of the few exceptions to the dichloro rule is 2,4-dichlorocinnamamide (figure 7.18), which does not crystallise with a 4 Å axis and does not photodimerise.

There is at present no fully satisfactory explanation for the dichloro rule. Although dispersion forces between chlorine atoms of adjacent molecules are expected to be large, the electronegativity of chlorine would favour

7.17.

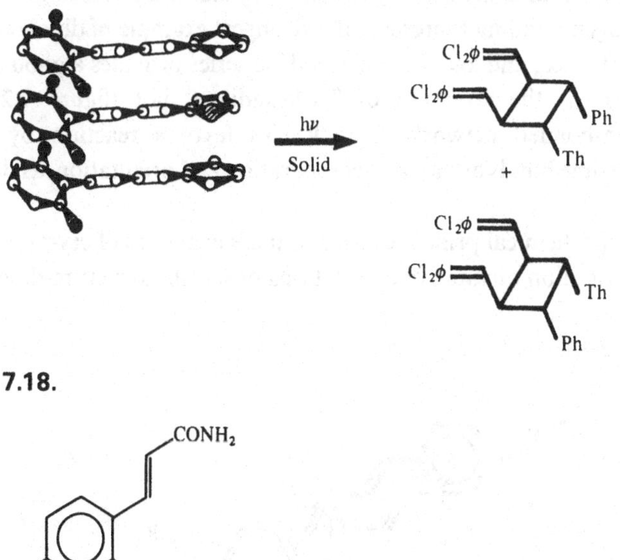

7.18.

repulsion between chlorines of neighbouring molecules if isotropic atom–atom potentials are assumed. It is generally considered that the introduction of the chlorine atoms leads to additional attractive forces which produce the desired short unit-cell dimension. As in the case of chlorine itself, it is likely that the anisotropic effect of the chlorine lone-pair electron density on the intermolecular potential must be taken into account to explain these structural features, as discussed in chapter 2.

The use of polar substituents to control orientation and hence reactivity is a more elegant approach to crystal engineering. This may either enhance reactivity or produce increased resistance to photodimerisation or polymerisation, depending on the system. For example, the polymerisation of crystalline phenylene diacrylic acids discussed earlier in this chapter is believed to be aided by the influence of the interaction between a carbonyl group of one molecule and the aromatic ring of a neighbour on the mutual orientation of the molecules. This orientation is as shown in figure 7.19 and the overlap of the carbonyl group and aromatic ring, which is reminiscent of that found in many quinone structures (chapter 3), results in the two double bonds involved in the reaction being parallel and close to each other. In contrast, in the *p*-toluenesulphonate derivative of diacetylene (PTS), replacement of the two methyl groups by chlorines (of similar size but conferring polarity on the ring-substituent bond) has the effect of steering neighbouring molecules into a configuration in which the reactive carbon atoms are too far apart for polymerisation, yielding a photostable material.

Hydrogen bonding represents the strongest example of dipolar orientational influence, and again the diacetylene series provides a good example as shown in the structure of 2,4-hexadiyne diol (figure 7.20). The hydrogen-bonded network of molecules favours reaction by placing reactive triple bonds at the correct separation and orientation, as described earlier.

The topochemical principles used in these examples of crystal engineering are based on qualitative correlations of specific structure-determining

7.19.

interactions and their influence on reactivity. The availability of computerised crystallographic databases, coupled with developments in the atom–atom potential method, has recently made it possible to analyse such correlations in a more quantitative way. In this approach[22], trends in the packing characteristics of a given series of compounds of known crystal structures which correlate with solid-state reactivity are first identified. Atom–atom potential calculations are then carried out to determine the interaction energy between a molecule and each of its successive nearest neighbours. These identify the size of the cluster of molecules whose interactions with the original molecule are dominant, and the pattern of interactions within this cluster (e.g. is the strongest interaction with one single nearest-neighbour molecule, or with two equally strongly interacting molecules, etc.?). Finally, the scope for chemically modifying the molecule (e.g. by substitution of hydrogen atoms by other groups), without destroying the important structural features conducive to reaction, is investigated by determining the tightness of the lattice packing at each

7.20.

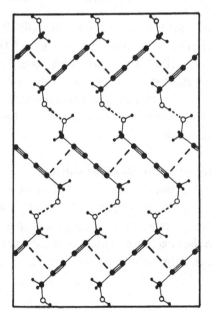

● = Carbon
o = Oxygen
• = Hydrogen
···· = Hydrogen bond
— — — = Bond formed on polymerisation

potential site for chemical modification. This is done conveniently by specifying a parameter known as the local packing density (n_i), defined by

$$n_i = \sum_{j=1,p} r_{ij}^{-2} \qquad (7.2)$$

where p is an integer specifying the maximum number of atoms, j, to be regarded as 'local' to the atom i. The higher the value of this packing density at a particular site, the less likely it is that substitution at this site will preserve the reactive structure. From known structural and reactivity data it is possible to specify critical values of n_i for each substituent, above which the favourable structure is disrupted. For example, Thomas and co-workers[22] analysed 15 known structures of 2-benzyl-5-benzylidene cyclopentanone (BBCP) (figure 7.21) and its derivatives and identified two types of packing, both involving clusters of six nearest molecules $(a-f)$ with interaction energies in the following order:

Type 1 $a>b>c=d>e=f$ Reactive (dimerise)
Type 2 $a=b>c=d>e=f$ Non-reactive

The local packing densities around the 18 hydrogen atoms in BBCP ranged from 0.72 to 3.06 and the critical values for replacement by Cl or CH_3 or by Br were in the ranges 1.85–2.49 and 0.72–1.85, respectively. As in the earlier example of diacetylene polymerisation, the structural limits for reactivity are defined more precisely as more structural and reactivity data become available, but these limits are specified in greater detail in the above approach than, for example, in figure 7.10.

7.5   Control of crystal growth and dissolution

Similar topochemical principles may be applied to processes occurring at crystal surfaces, for example adsorption, crystal growth and crystal dissolution[23]. Thus, for example, it is frequently found that crystals develop in ways which reflect the direction of strongest interaction between neighbouring molecules. In electron donor–acceptor complexes, charge-

7.21.

transfer interactions are strongest along the heterosoric stacks and very frequently crystals of such complexes grow as needles elongated along this stacking direction. Using this principle, it would also be expected that the occurrence of significant inter-stack interactions in charge-transfer salts should be reflected in the formation of plate-like crystals, or at least needles with well-developed faces, rather than very thin elongated needles. Conversely, adsorption onto a particular crystal face of an impurity which is topologically compatible with that crystal face, but which presents a reverse side onto which further molecules of the host material cannot adsorb, will inhibit growth perpendicular to that face. If the impurity is carefully selected to be topologically compatible only with this face, the result will be enlargement of the face as the crystal continues to grow along all other directions except perpendicular to the face. Thus in figure 7.22, benzamide crystallises from ethanol as {001} plate crystals elongated along *b*, but addition of benzoic acid to the crystallising solution selectively inhibits growth along the *b*-direction, yielding crystals elongated along *a*.

7.22.

Atom–atom potential calculations of the preferred adsorption site for benzoic acid on benzamide confirm that this is on the {001} crystal faces. The retardation of growth rate along b is the result of replacement of an attractive N–H $\cdots$ O hydrogen bond by a repulsive O $\cdots$ O interaction, with a calculated energy loss of over $40 \, kJ \, mol^{-1}$. Inhibition using o-toluamide gives crystals elongated along b, while p-toluamide inhibits growth along the c-direction, giving thinner plates, and both of these effects are also consistent with the predicted preferred mode of adsorption of the inhibitor.

Inhibition of crystal growth has also been used as a method for resolution of racemic mixtures. Addition of a small amount of a chiral inhibitor selectively inhibits growth of crystals of the enantiomer having the same configuration as the additive, so that crystals of the other enantiomer form first and can be separated. In favourable cases, 100% resolution has been achieved by this method; for example, (RS)-glutamic acid$\cdot$HCl$+(S)$-lysine; (RS)-threonine$+(S)$-glutamic acid; and (RS)-asparagine$+(S)$-aspartic acid, in all cases with the R enantiomer precipitating first. In a related approach, resolution is achieved by selective adsorption at an interface. Thus, glycine {001} plates tend to float at air–water interfaces with either (010) or (0$\bar{1}$0) pointing towards the water, and in aqueous solution of a racemic mixture, crystals with the (010) face down will only attract the (R)-acid molecules while conversely crystals with the (0$\bar{1}$0) face down will attract only the (S)-acids. By using the underside of a Langmuir–Blodgett film of a resolved amino acid with a hydrophobic side chain (e.g. (S)-α-aminooctanoic acid) as a template, it is possible to ensure that all crystals at the interface have the same orientation, thus generating optically pure crystals from a mixed solution. In addition to its elegance, this approach provides one possible explanation of the origin of natural optical activity, as the first crystals of amino acids formed in the 'pre-biotic soup' and floated up to interfaces.

Finally, selective etching of specific faces of crystals can be achieved using the same principles of topochemically preferred adsorption, together with the idea that growth inhibitors also inhibit dissolution. Thus, selective adsorption of the inhibitor onto one face protects the normal regions of that face from dissolution and enhances the contrast between these regions and the sites of emerging dislocations, where other facial surfaces not protected by the inhibitor are exposed and dissolve in the etchant. For example, glycine {010} plates have well-developed etch pits generated only on their (010) faces by dilute solutions of glycine containing (R)-alanine as an inhibitor, while etching with (S)-alanine as inhibitor only affects the (0$\bar{1}$0) faces. This selective etching can serve to identify crystal faces and should find wide application.

7.6 The role of defects

For a substantial number of molecular crystals, the products of photochemical reactions are not those predicted from the application of established topochemical principles in conjunction with the known crystal structure, although the reactions are still stereospecific, suggesting that lattice-control is still active[24]. One example which has already been discussed in chapter 4 is the photodimerisation of 9-cyanoanthracene, where the crystal structure would predict the *cis* product although in practice the *trans* product is obtained. As shown in figure 4.6 (p. 57), stacking faults can readily be visualised which would bring significant numbers of molecular pairs into the required *trans* registry at the defect plane. Similarly, there are molecular crystals whose structures do not bring neighbouring molecules close enough together for any photochemical reaction to be likely on topochemical grounds, yet which nevertheless do undergo such reactions. One such example is anthracene itself, where calculations using atom–atom potentials have shown[25] that there are at least three structures close in energy to the observed phase yet of differing structure in that they produce relative orientations of neighbouring anthracene molecules which are favourable to dimerisation. As mentioned in chapter 3, electron diffraction using electron microscope equipment (see also chapter 4) has confirmed the existence of small islands of one of these new phases (triclinic) in strained crystals of anthracene[26].

These observations, although they point towards possible mechanisms of such 'topologically abnormal' reactions, raise the question of why defects should dominate the course of the reaction even though their bulk concentrations are very much smaller than those for normally oriented molecules. The answer to this question is clear from the description of the nature of electronic excitation in crystals given in chapter 6. The excitation is mobile in the form of excitons, which are trapped on molecules at defect sites whose energies are slightly different from those of molecules in normal lattice sites. The longer residence time for the excitation at such trap sites increases the probability of reaction at these sites. Furthermore, at structural defects there is frequently more space for molecules to move into correct registry for reaction, as well as more space to accommodate the differently shaped product molecule. Evidence supporting this exciton mechanism is obtained from experiments incorporating fluorescent dopants into crystals of 9-cyanoanthracene. Increasing dopant levels increased fluorescence yield directly in proportion with a diminution of the photodimer yield, as expected for introduction of a process in direct competition as an exciton trap. Also, the quantum yield for the photoreaction was found to increase as the reaction progressed, consistent with reaction originating at defects, whose number proliferates as the differently shaped product molecules replace the host structure[27].

These examples show that the important structural feature in topochemical reactions is the local site cage around the reactive site in the solid, and that this site need not necessarily be the normal lattice site. This concept of the local site cage is particularly important for reactions of impurity guest molecules in crystals. Using atom–atom potentials it is now possible to calculate the preferred orientation of a given impurity molecule in a host lattice. In several cases so far studied, the predicted orientation is not one that is conducive to reaction, yet reactions have been observed. This can be explained only if the impurity preferentially occupies disordered regions around defects (as discussed in chapter 4) or if the local structure changes on electronic excitation. The latter possibility is not unrealistic, because an electronically excited molecule is generally more polarisable than the ground-state molecule and as such will have bigger dispersion forces attracting it to its neighbours. Therefore, it may have a tendency to move off its normal lattice site if the excitation remains localised for long enough, and the conversion of excited-state molecules to reaction centres may then become coupled with lattice vibrations (phonons), with localised dynamic disorder at reaction sites.

The quantitative modelling of such effects is extremely complex and, although they can be used to rationalise anomalous topological effects, it remains difficult if not impossible to predict which systems will show normal behaviour and which will be dominated by defects. Since it cannot be claimed that all crystals showing normal topochemical reactions are free from defects, nor that electronic excitation is mobile only in crystals where defects dominate, this remains a challenging problem for both experimental and theoretical research.

The examples in this chapter have shown the wide scope of chemical reactions in molecular crystals. The clear advantages of carrying out organic reactions in the solid state are improved specificity, minimisation of impurities and maximisation of yields by reducing the chance of side-reactions, and in some cases the formation of materials otherwise unobtainable (e.g. single-crystal polydiacetylenes). These advantages, together with its scientifically stimulating multidisciplinary nature, suggest that this area of solid-state chemistry is likely to develop still further in future despite the fact that the reactions are by their nature limited to fairly small quantities of material and as such are not readily adaptable to large-scale plant in the chemical industry.

References

1 C.H. Bamford and G.C. Eastmond, *Quart. Rev. (London)*, 1969, **23**, 271.

2 G. Adler and W. Reams, *J. Chem. Phys.*, 1960, **32**, 1698.

3 I.F. Brown and D.M. White, *J. Am. Chem. Soc.*, 1960, **82**, 5671.

4 G. Friedman, M. Lahav and G.M.J. Schmidt, *J. Chem. Soc.*, *Perkin II*, 1974, 428.

5 G. Quinkert, T. Tabata, E.A.J. Hickman and W. Dobrat, *Angew. Chem. (Int. Edn)*, 1971, **10**, 199.

6 W. Heggie and J.K. Sutherland, *Chem. Commun.*, 1972, 957.

7 C. Liebermann, *Chem. Ber.*, 1889, **22**, 124, 782.

8 G.M.J. Schmidt, *Pure Appl. Chem.*, 1971, **27**, 647.

9 M.D. Cohen and B.S. Green, *Chem. Brit.*, 1973, **9**, 490.

10 A.A. Dzakpasu, S.E. Phillips, J.R. Scheffer and J. Trotter, *J. Am. Chem. Soc.*, 1976, **98**, 6049.

11 H. Nakanishi, M. Hasegawa and Y. Sasada, *J. Polymer Sci.*, A-2, 1972, **10**, 1537.

12 H. Nakanishi, M. Hasegawa and Y. Sasada, *J. Polymer Sci.*, *Polym. Phys. Ed.*, 1977, **15**, 173.

13 G.N. Gerasimov, O.B. Mikova, E.B. Kotin, N.S. Nekhoroshev and A.D. Abkin, *Dokl. Akad. Nauk. SSSR*, 1974, **216**, 1051.

14 G. Wegner, *Z. Naturforsch.*, 1969, **24b**, 824.

15 D. Bloor, *Mol. Cryst. Liq. Cryst.*, 1983, **93**, 183.

16 G.M.J. Schmidt, in *Reactivity of the Photoexcited Organic Molecule*, New York: Wiley, 1967, p. 227.

17 D. Bloor in *Quantum Chemistry of Polymers – Solid State Aspects*, ed. J. Ladik, Dordrecht: Reidel, 1984, p. 191.

18 D. Bloor, *Phil. Trans. Roy. Soc. Lond.*, 1985, **A314**, 51.

19 H. Sixl, in *Polydiacetylenes*, Proceedings of a NATO Advanced Research Workshop, Dordrecht: Nijdorf, 1984.

20 J. Bregman, K. Osaki, G.M.J. Schmidt and F.I. Sonntag, *J. Chem. Soc.*, 1964, 2021.

21 J.A.R.P. Sarma and G.R. Desiraju, *Acc. Chem. Res.*, 1986, **19**, 222.

22 N.W. Thomas, S. Ramdas and J.M. Thomas, *Proc. Roy. Soc. Lond.*, 1985, **A400**, 219.

23 L. Addadi, Z. Berkovitch-Yellin, I. Weissbuch, J. van Mil, L.J.W. Shimon, M. Lahav and L. Leiserowitz, *Angew. Chem. (Int. Edn)*, 1985, **24**, 466.

24 J.M. Thomas, S.E. Morsi and J.P. Desvergne, *Adv. Phys. Org. Chem.*, 1977, **15**, 63.

25 D.P. Craig, J.F. Ogilvie and P.A. Reynolds, *J. Chem. Soc. Faraday II*, 1976, **72**, 1603.

26 W. Jones and J.M. Thomas, *Prog. Solid State Chem.*, 1979, **12**, 101.

27 M.D. Cohen, in *Reactivity of Solids*, 7th Int. Symp., Bristol, ed. J.S. Anderson, M.W. Roberts and F.S. Stone, London: Chapman and Hall, 1972, p. 402.

8

Electrical properties

Organic solids are commonly regarded as electrical insulators since they frequently contain few charged species and exhibit poor overlap between orbitals of neighbouring molecules so that charge cannot pass rapidly from molecule to molecule. Although it is true that organic materials (e.g. plastics) are among the most widely used electrical insulators, the electrical properties of molecular crystals are in fact a great deal more varied and interesting than the above view would suggest.

As shown in figure 8.1, the range of electrical conductivities observed to date in organic materials is as wide as that for inorganic materials, and organic metals and organic superconductors are now a reality. As early as 1906 Pocchetino[1] discovered that anthracene was a photoconductor, but the area remained remarkably unexplored until 1948, notwithstanding the suggestion by Szent Györgyi in 1941 that electronic conductivity in organic materials might be important in biological systems. Then work by Eley in Britain[2] and Vartanyan in Russia[3] on semiconductivity of phthalo-cyanines, and by Akamatu in Japan[4] on electrical properties of polycyclic aromatic hydrocarbons marked the beginning of the systematic study of organic semiconductors. Since the early 1970s this field has developed from relative obscurity to one of the most active areas of solid-state physics and chemistry. This development has been fuelled by the goal of combining the imaginative diversity of synthetic organic chemistry with the spectacular growth of micro-electronics and its constant search for new materials.

8.1 Charge-carrier mobility

The electrical conductivity of a material is determined by the number, n, of charge carriers of charge $|e|$ and mobility μ:

$$\sigma = n|e|\mu. \tag{8.1}$$

The charge carriers may be electrons, with negative charge, or holes, with positive charge. Their mobility may be determined by the very direct method of measuring the time taken for a thin sheet of charge carriers (thickness δ) generated on one crystal face to traverse the crystal (thickness d) under the influence of a known applied electric field[5] (hence the units of mobility are $cm^2\,V^{-1}\,s^{-1}$, i.e. carrier velocity per unit field gradient). The sheet of charge carriers is most commonly generated by a fast pulse (duration much shorter than the transit time, t, of charge carriers through the crystal) of light of a wavelength strongly absorbed by the sample, illuminating a semi-transparent electrode on one crystal face (figure 8.2). The current is measured with a fast-response FET amplifier measuring the voltage drop ΔV across resistor R. Depending on the time constant CR, the system will measure the duration of an (ideally) steady current pulse (figure 8.2(*a*)) or the total integrated current (figure 8.2(*b*)).

In practice the transit time may not be as well defined as implied by figure 8.2(*a*) and (*b*), since imperfections in real crystals result in a spreading out of the initially thin sheet of charge carriers as some carriers encounter traps or

8.1.

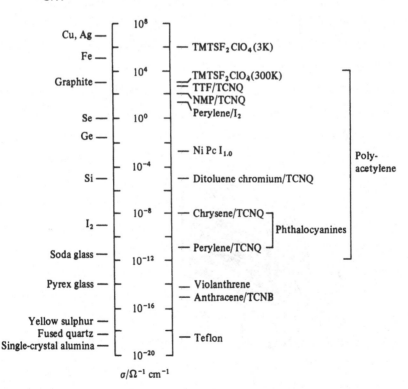

$\sigma/\Omega^{-1}\,cm^{-1}$

defects. This is particularly problematic at low temperatures, where the number of traps whose depth is large compared with kT is greatly increased. In this case it would be preferable to observe directly the arrival of charge at the reverse side of the crystal; this can be done by replacing the single-collector electrode with an interdigitated electrode[6] (figure 8.3). A

8.2.

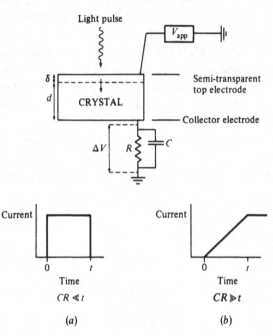

(a)   (b)

8.3.

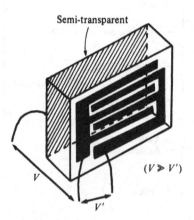

relatively large potential is applied between this and the semi-transparent electrode, while a smaller potential difference is maintained between the two halves of the interdigitated electrode. During transit of the charge sheet from the semi-transparent electrode across the crystal, the current flowing between the two halves of the split electrode is negligible. As the charge sheet arrives at the split electrode, the current between the two halves rises as a result of the increased carrier concentration in the gap. Measuring this current as a function of time thus produces a peak whose shape reflects the distribution of arrival times. Two practical problems of this method are the difficulty of ensuring good contact between the interdigitated split electrode and the crystal surface, and that of obtaining very small interdigitated electrode assemblies for use with small crystals.

Table 8.1 summarises some of the data obtained for carrier mobilities in molecular crystals using these methods. Room-temperature mobilities are typically of the order of $1\,\text{cm}^2\,\text{V}^{-1}\,\text{s}^{-1}$, which is a factor of approximately 10^4 smaller than those in typical inorganic semiconductors such as silicon or germanium. They are also anisotropic, as is their temperature-dependence, which is to be expected in view of the structural anisotropy of most molecular crystals (chapter 3). The low mobility is a consequence of the poor overlap between adjacent molecules in crystals. Three principal types of theoretical model have been used to describe charge-carrier mobility in molecular crystals: the band model; the hopping model; and the tunnelling model.

8.2   The band model

In the band model, intermolecular interaction leads to the formation of bands of energy states for electrons and holes. This occurs by removal of the degeneracy of molecular electron–hole states, in the same way as already described in chapter 6 for the formation of exciton bands in crystals. These bands are in effect delocalised states in which electrons and holes can move from molecule to molecule in a periodic lattice. The basic ideas of band theory as applied to metals and inorganic semiconductors are extensively covered in solid-state physics[7] and chemistry[8] texts at a variety of levels, so at this point we shall concentrate on the features specific to the application of band theory to molecular crystals.

The poor overlap between adjacent molecules in crystals results in the bandwidth becoming narrow. If the bandwidth becomes comparable to kT (0.026 eV at room temperature) it follows that all the levels in the band can be thermally populated rather than merely the levels at the lower end of the band. In this case, the approximations behind the effective-mass concept

Table 8.1 *Charge-carrier mobilities*

Crystal	Carrier	Direction	Mobility/ cm^2 V^{-1} s^{-1} ($T = 300$ K)	$\mu \alpha T^n$ n	Measurement range/K
Benzene	−	A	1.5 (287 K)	−2.0	173–278
p-Diiodobenzene	+	A	12	−0.5	240–310
	+	B	4	0	240–310
	+	C	2	−0.8	240–310
Naphthalene	−	A	0.51	−0.1	220–300
	−	B	0.63	0	220–300
	−	C′	0.68	−0.9	220–300
	+	A	0.88	−1.0	220–300
	+	B	1.41	−0.8	220–300
	+	C′	0.5, 0.99	−2.1	220–300
Azulene	−		0.15	0	220–360
1,4-Dibromo-naphthalene	−	A	0.017	≈ −2	270–300
	−	B	0.013	≈ −2	270–300
	−	C′	0.034	≈ −2	270–300
	+	A	0.66	≈ −2	270–300
	+	B	0.25	≈ −2	270–300
	+	C′	0.87	≈ −2	270–300
Anthracene	−	A	1.6	−1.0	77–300
	−	B	1.0	−0.2, −0.6	170–380
	−	C′	0.4	+0.8, 0	80–450
	+	A	1.2	−1.0	300–400
	+	B	2.0	−1.0	300–400
	+	C′	0.8	−1.0	170–450
Perdeuterated anthracene	−	A	1.7	−1.8	280–400
	−	B	0.99	−1.4	280–400
	−	C′	0.35	0	280–400
	+	A	1.1	−1.7	280–400
	+	B	2.0	−1.4	280–400
	+	C′	0.78	−1.0	280–400
Phenazine	−	A	0.29	0	180–360
	−	B	1.1	−0.65	230–360
	−	C′	0.5	−0.1	230–360
Phenothiazine	−	∥1.*	1.7	−3	300–350
	+	⊥*	5.0	−3	300–400
N-isopropyl-carbazole	−	C	1.0	0	244–370
Pyrene	−	A–B	0.7	−1.5	260–350
	−	C′	0.5	−2.0	260–350
	+	A–B	0.7	−1.6	260–350
	+	C′	0.5	−1.3	260–350
β-Phthalocyanine	+	C	1.1	−1.3	290–600
	−	C	1.4	−1.5	290–600
TCNQ	−	⊥(001)	0.4	0	204–306
	+	⊥(001)	0.4	0	204–306
S$_8$	−	all	10^{-4}	$exp(-0.17/kT)$	220–413
	+	⊥(111)	10	−1.5	300–400
Se$_8$	−	⊥(100)	2	−1.5	300–400

used in conventional band theory are no longer valid and the mobility is given by[9]

$$\mu = (e/kT) < \tau_r v_i^2 > \tag{8.2}$$

where τ_r is a relaxation time and v_i is the electron velocity in the ith direction, the angular brackets indicating thermal averaging. The velocity can be related to the intermolecular electron-overlap integral (J) and the lattice constant a by

$$v_i = Ja/\hbar. \tag{8.3}$$

Hence,

$$\mu = (e\tau_r/kT)(J^2 a^2/\hbar^2) = (e\lambda/kT)(Ja/\hbar) \tag{8.4}$$

where

$$\lambda = v_i \tau_r = \text{mean free path of the electron.} \tag{8.5}$$

As mentioned above, μ can be measured experimentally and the lattice constant a is known from X-ray diffraction data, while the overlap integral J can be estimated theoretically. Hence equation 8.4 can be used to estimate the mean free path of the charge carriers[10]. For anthracene, J has been estimated to be of the order of 0.01 eV, so that by using $\mu = 10^{-4}\, \text{m}^2\,\text{V}^{-1}\,\text{s}^{-1}$ and $a = 6 \times 10^{-10}\,\text{m}$ we have

$$\lambda = \mu kT\hbar/eJa = (10^{-4} \times 0.026 \times 6.58 \times 10^{-16})/(0.01 \times 6 \times 10^{-10})$$
$$= 2.9 \times 10^{-10}\,\text{m}.$$

This value is of the same order as the intermolecular spacing in anthracene, and it is clearly not realistic to describe carrier motion in terms of a delocalised band model if this is the case. Since the above argument is open to criticism on the grounds that it uses theoretical estimates for J which may be in error, a better test of the validity of the band model to describe charge-carrier motion in molecular crystals is to compare the temperature-dependence of mobility with the predictions of narrow-band theory. From equation 8.4 it can be seen that mobility should at least vary as T^{-1}, plus any additional variation resulting from the temperature-dependence of a, J and λ. The variation of lattice constant a with temperature (thermal expansivity) is small, with

$$a = a_0(1 + \beta T) \tag{8.6}$$

where β is the coefficient of thermal expansion (typically $10^{-4}\,\text{K}^{-1}$). The electron-overlap integral J decreases approximately exponentially as the lattice expands, but with the small expansivity this effect amounts only to around 5% over a 200 K range. Furthermore, since electron-overlap integrals are involved in all theories of charge transport, this component of the temperature-dependence is not diagnostic for a band model.

The variation of mean free path λ, or relaxation time τ_r, with temperature depends on the dominant factor limiting mean free path. In the simplest case of charged or neutral impurities limiting carrier-free path, no additional temperature-dependence is involved. In perfect crystals where the mean free path is not limited by impurities, scattering by lattice vibrations is the limiting factor and electron–phonon coupling becomes important. There are two origins of electron–phonon coupling. First, electron overlap integrals, J, depend on intermolecular separation and hence on lattice vibrations. Since the overlap integrals are determined by overlap of the tails of molecule and molecular-ion wave-functions at a distance where these are falling off quite steeply, this dependence on lattice vibrations is expected to be quite strong. Second, charge-carrier energies in molecular crystals depend on the polarisation energies of charged species in the crystal. These polarisation energies, as the name would suggest, arise from polarisation of the molecules surrounding each charged species in the lattice and are the solid-state counterpart of solvation energies of ions in solution. Methods for calculating and measuring polarisation energies will be discussed later in this chapter, but at this stage it suffices to know that they vary as r^{-4}, where r is the intermolecular separation (hence they are dependent on lattice vibrations), and that they are an order of magnitude greater than the energy required to move a charge carrier from one molecule to another (so their dependence on lattice vibrations is quantitatively more important than that of the electron-overlap integral).

As described in chapter 5, lattice modes may be categorised as optical or acoustic phonons, and both may couple with electrons, although optical phonons generally involve larger displacements of adjacent molecules relative to each other and so may be expected to have a larger influence on charge-transport properties. The coupling is generally considered to first-order only (i.e. using a relationship between the positions of the molecules and the electron energy) although second-order (quadratic) coupling (involving a relationship between the electron energy and the force constants between molecules) has also been used in refined models. In some cases molecular vibrations, as well as lattice vibrations, may couple with the electrons, particularly where the equilibrium geometry of a molecule changes significantly on ionisation, although normally such coupling is much smaller than that with lattice vibrations[11].

Table 8.2 shows the calculated temperature-dependence of mean free path λ and mobility μ deduced from first-order coupling of acoustic and optical phonons in a narrow-band model[10]. Even though hv and kT are comparable at room temperature for optical phonons in molecular crystals, so that $(e^{hv/kT}-1)$ can be quite well approximated as $[1+(hv/kT)]-1$, i.e. hv/kT, these temperature-dependences are still substantial and provide a good basis for analysis of experimental data on mobility as a function of temperature. Unfortunately, examination of such data provides examples consistent with electrons coupling with acoustic or optical phonons, depending on the system, while yet other systems show temperature-independent mobility in certain directions (see table 8.1). Durene (1,2,4,5-tetramethylbenzene) is one example where both electron and hole mobilities show strong temperature-dependences between T^{-2} and T^{-3} as required by acoustic–phonon coupling (figure 8.4)[12], whereas the electron

Table 8.2 *Temperature-dependence of charge-carrier mean free path* (λ) *and mobility* (μ) *according to the narrow-band model with first-order phonon coupling*

Phonon		λ	μ
Acoustic,	1 phonon	T^{-1}	T^{-2}
	2 phonon	T^{-2}	T^{-3}
Optical,	1 phonon	$T[\exp(hv/kT)-1]$	$\exp(hv/kT)-1$
	2 phonon	$T[\exp(hv/kT)-1]^2$	$[\exp(hv/kT)-1]^2$

8.4. Charge-carrier mobility in durene.

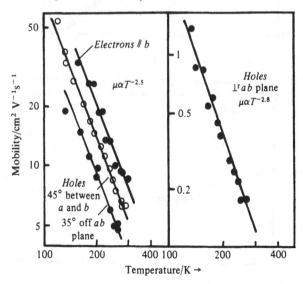

mobility in the *a*-direction of naphthalene (figure 8.5)[13] has a much smaller temperature-dependence of $T^{-1.53}$, as expected for optical–phonon coupling. In the latter case the data could be fitted well using just two vibration frequencies, an optical phonon of frequency 67 cm^{-1} and an intramolecular vibration of frequency 512 cm^{-1}, and the effect of deuteration on mobility could also be well modelled by its effect in reducing these two frequencies.

More complex behaviour has been observed for electron mobility in the *c′*-direction[14] as well as for hole mobility in various directions[15] in naphthalene, as shown in figures 8.6 and 8.7, respectively. At low temperature, below about 100 K, the electron mobility in the *c′*-direction rises, whereas above about 100 K it is temperature-independent. (In the

8.5.

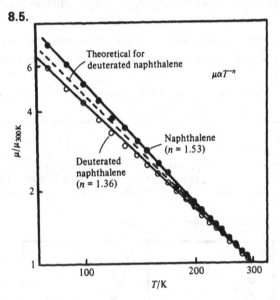

8.6. Electron mobility in the *C′*-direction in naphthalene.

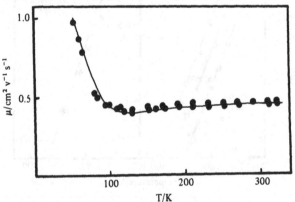

equivalent direction in anthracene this temperature-independent electron mobility extends from 100 to nearly 500 K.) Although no simple band model can account for this temperature-independent mobility, the region below 100 K is well fitted[16] by a narrow-band model with scattering by optical phonons of frequency 53 cm^{-1}. Unfortunately, there are not many molecular crystals that can be obtained pure enough to permit mobility measurements at such low temperatures, where even normally-shallow traps become deep compared with kT. However, Warta and Karl[17] were able to measure drift mobilities showing essentially no influence of deep trapping in an ultrapure sample of naphthalene down to 4.2 K for holes and 23 K for electrons along the a-direction, and they showed that the $T^{-2.9}$ dependence of mobility continues to approximately 40 K. At this point the mobility is of the order of $100 \text{ cm}^2 \text{ V}^{-1} \text{ s}^{-1}$, and further cooling leads to an apparent levelling-off of the mobility (shown in figure 8.7) corresponding to a limiting drift velocity of $1.8 \times 10^6 \text{ cm s}^{-1}$. The apparent drift mobility varies with the applied electric field, as shown in figure 8.7, but when the actual drift velocities are plotted as a function of applied electric field it is clear that this effect is in fact a saturation of drift velocity (figure 8.8). These observations have been interpreted in terms of standard wide-band semiconductor theory, according to which the measured mobility of $240 \text{ cm}^2 \text{ V}^{-1} \text{ s}^{-1}$ is a lower limit to the true Ohmic mobility and corresponds to a mean free path of at least eight lattice constants. The standard

8.7. Charge-carrier mobility in ultrapure naphthalene.

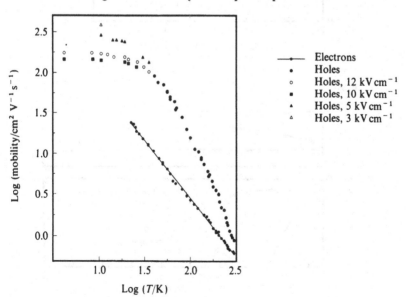

band model thus appears entirely reasonable in these extreme conditions of low temperature and ultra-high purity.

The importance of using high-quality pure crystals if meaningful results are to be obtained from drift mobility experiments is further emphasised by experiments in which hole and electron mobilities were measured for the c'-direction of anthracene for high-purity crystals and crystals doped with 10^{-7} mol/mol tetracene (figure 8.9)[18]. These results show that, even at these low levels, trapping by tetracene is dominant at the low-temperature end of the measured range and that tetracene is a more effective trap for holes than for electrons. The low-temperature limiting slopes of the $\log \mu$

8.8. Electric-field dependence of hole drift velocity in ultrapure naphthalene.

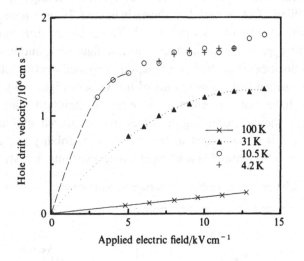

8.9.

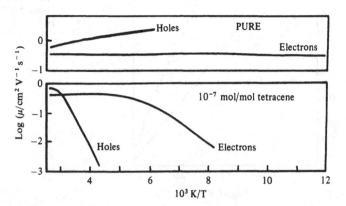

versus $1/T$ plots for the doped crystal provide information on these trap depths rather than on the inter-trap transport mechanism.

8.3 The hopping model

Implicit in the band model is the assumption that charge-carrier mean free paths extend over several lattice sites, with the residence time on any one site being short compared with the time that would be taken for the lattice to relax around a molecule which suddenly acquires an immobile charge. If the mean free path is of the order of the intermolecular distances, or if the charge-carrier motion is effectively a series of rapid jumps between trap sites where the carriers spend most of their time, this assumption is no longer valid. The carriers in such a case polarise the surrounding lattice, and both the molecules on which the charges are localised and their nearest neighbour environments relax their structure to new equilibrium positions. The charge carrier and its associated deformation form a quasi-particle known as a polaron. In the case of most molecular crystals the deformation is localised to nearest neighbour molecules and the system is known as a small polaron. At the equilibrium position the stabilisation energy associated with the polaron is maximised. Any movement of the charge from its equilibrium position decreases the stabilisation energy. Thus, charge carriers essentially reside in potential wells centred at the equilibrium positions. To move from one site to a neighbouring site, the polaron must traverse a barrier as shown in figure 8.10. Charge-carrier motion is thus a series of 'hops' from one site to the next, hence the term 'hopping model'.

The probability of such hops taking place is calculated as the product of the probability of the carrier achieving energy E' and the probability of a carrier of this energy undergoing transfer to the neighbour site. It is

8.10.

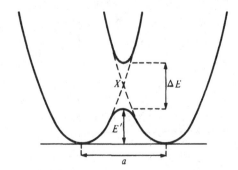

proportional to $T^{-m}\exp(-E'/kT)$, with m typically 0 or $\frac{1}{2}$ depending on conditions[19]. In turn, the mobility is given by[20]

$$\mu \propto (ea^2/kT)T^{-m}\exp(-E'/kT),$$

i.e. $$\mu \propto T^{-n}\exp(-E'/kT) \qquad (8.7)$$

with $n = 1$–1.5. The barrier height E' is approximately half the polaron binding energy less half the amount ΔE by which the degeneracy at the cross-over point X in figure 8.10 is lifted because of interaction between neighbouring molecules[21]. However, E' is clearly a function of a and as such provides an obvious link with phonons, which can be regarded as periodic variations of a.

At low temperatures the exponential term dominates equation 8.7 so that, in marked contrast to the predictions of band theory, mobility is expected to decrease sharply. In effect this means that at sufficiently low temperatures the dominant charge-transport mechanism in molecular crystals will be via the narrow-band model, as observed in naphthalene (figures 8.6 and 8.8). At high temperatures, particularly if the barrier height E' is low, the pre-exponential T^{-n} dependence dominates, and it is thus difficult to differentiate band and hopping transport from their temperature-dependence in this range. (Indeed, if kT is large compared with the barrier height, the carrier becomes effectively delocalised and hence its motion is best described in terms of a band model with scattering increasing with temperature.) Charge transport in sulphur crystals[22] exhibits a temperature-dependence following equation 8.7 (figure 8.11).

8.11. Charge-carrier mobility in sulphur (with shallow hole-trap densities/ml).

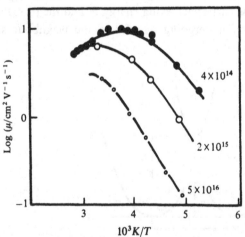

Temperature-independent mobility, as in the c'-direction of anthracene, is not adequately explained by either the band or hopping models, although Sumi[23] has proposed that transport in this direction is dominated by hopping coupled with librational modes of vibration which are very direction-dependent. This model is consistent with temperature-independent mobiility, but predicts a field-dependence at high electric fields[24] which is not observed[25]. Therefore, although these models account for many features of charge transport in molecular crystals, this area still presents challenges to theoreticians and experimentalists.

8.4 The tunnel model

Wherever models of charge transport are proposed which involve potential barriers, quantum-mechanical tunnelling of charge through the barrier may be invoked in addition to hopping models. Tunnelling depends on barrier thickness, being most likely for thin barriers, and hence is temperature-dependent via lattice vibrations, which in this context are regarded as periodic fluctuations in the thickness of the intermolecular potential barrier. This model was one of the earliest used for charge transport in molecular crystals and was developed by Eley and co-workers[26]. Since its predictions are in general similar to those of hopping models it is often not mentioned as a separate model, although quantitative comparisons of mobility calculations with experiment should always include a tunnelling contribution.

One problem common to both hopping and tunnelling models is their prediction that electrons, which occupy more energetic conduction states than the positive holes which they leave behind, should have higher mobility since the barrier to their motion is lower and thinner than that for holes. As can be seen (e.g. from table 8.1 and figures 8.6, 8.7 and 8.9) this is not generally observed. An advantage common to both models is that they provide a possible explanation for the Meyer–Neldel rule[27] or compensation law, which states that there is a relationship, within a given group of materials, between the semiconduction activation energy (E) and pre-exponential factor (σ_0) of the form

$$\ln\sigma_0 = \alpha E + \beta \tag{8.8}$$

where α and β are constants,

$$\sigma = \sigma_0 \exp(-E/kT) \tag{8.9}$$

and

$$\sigma_0 = Ne\mu \tag{8.10}$$

where N is an appropriate density of states.

The interpretation of equations 8.9 and 8.10 will be discussed later, but from equations 8.8–8.10 it can be seen that one description of the compensation law is that high mobility μ is associated with high charge-carrier creation energy E. To take the extent that high E implies charge carriers whose energy is nearer to the top of the intermolecular potential energy barrier E', on both the hopping and tunnelling models correspondingly high mobility μ is implied.

8.5   Thermal generation of charge carriers

So far this chapter has been concerned with the motion of charge carriers, and to understand the conductivity of molecular crystals we must now turn our attention to the factors affecting the number of these charge carriers. The distribution of both electrons and holes amongst the available energy levels in a molecular crystal is governed by the Fermi–Dirac function, according to which the probability $f(E)$ of an electron occupying an energy level E is given by

$$f(E) = \{1 + \exp[(E - E_F)/kT]\}^{-1}. \tag{8.11}$$

The energy E_F is known as the Fermi level, and is the energy of the state whose probability of occupancy is 0.5. It is in effect the chemical potential of electrons in the solid, and it is important to realise that there need be no actual electrons of precisely this energy in the crystal: it is merely a statistical average. For energy levels several times kT greater than E_F, the Fermi–Dirac function approximates to the classical Maxwell–Boltzmann distribution

$$f(E) = \exp[(E_F - E)/kT]. \tag{8.12}$$

The probability of a hole occupying the same energy level E is naturally given by $[1 - f(E)]$,

i.e. $\exp[(E - E_F)/kT]. \tag{8.13}$

Without necessarily implying that the band model is valid for molecular crystals, it is useful to describe the energy levels of electrons in terms of valence band and conduction band in the case of intrinsic conduction. Figure 8.12 is a simple energy-level diagram for the formation of separated charge carriers in a molecular crystal. The energy gap is given by[28]

$$\Delta E = IP - EA - P^+ - P^- + \Delta W_f \tag{8.14}$$

where P^+ and P^- are polarisation energies and ΔW_f is the stabilisation of the ground state on going from gas phase to crystal.

The number of holes and electrons produced at temperature T is given by the product of the density of states in the valence or conduction band and the probability of occupying the levels at E_v and E_c. For intrinsic conduction in a molecular crystal, the relevant density of states is approximately the number of molecules per unit volume, N, and since the number of holes and electrons must be equal for electrical neutrality,

$$N \exp[(E_F - E_c)/kT] = N \exp[(E_v - E_F)/kt]. \tag{8.15}$$

Hence

$$E_F = (E_v + E_c)/2 \tag{8.16}$$

and the Fermi level is located mid-way between valence and conduction bands. Thus, the carrier concentration n is given by

$$n = N \exp(-\Delta E/2kT). \tag{8.17}$$

It is possible to estimate the value of ΔE for many molecular crystals using equation 8.14, since values of gas-phase ionisation potential and electron affinity have been determined experimentally for many molecules and the term $(P^+ + P^- - \Delta W_f)$ may either be calculated or deduced directly from comparison of gas-phase and solid-phase ionisation potential data. The calculation of polarisation energies is in principle very complex, as ideally the charge distributions on the molecular ions should be calculated and then the interaction between these distributed charges and the anisotropic local environment should be estimated. This would require knowledge of local dielectric constants and anisotropic polarisabilities as well as some allowance for the effects of molecular and lattice vibrations. However, useful progress has been made using a variety of simplifying

8.12.

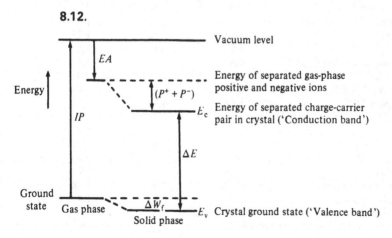

assumptions[29], and the influence of lattice defects and surfaces on polarisation energies has also been considered[30]. Nevertheless, the bulk of the available data on polarisation energies comes from the experimental approach; table 8.3 shows some of the values which have been obtained[31].

An immediately obvious and useful feature of these results is the similarity of the values for a wide range of materials. This result, which arises because larger molecules have higher polarisability yet lower polarising power when charged so that the interaction of polarisable and polarising species remains almost constant irrespective of size, means that for most materials the error involved in assuming $(P^+ + P^- - \Delta W_f) \approx 3.5 \, \text{eV}$ will be less than about $0.5 \, \text{eV}$. This is usually close enough to use equation 8.14 to find ΔE and compare it with the observed activation energy E. If $E \approx \Delta E/2$, intrinsic conduction is a possibility, although extrinsic processes giving comparable activation energies can only be safely eliminated if both E and σ_0 are determined. Since the relevant density of states will be much smaller for impurity-dominated

Table 8.3 *Experimental polarisation energies for positive charge carriers*

Compound	I_G/eV	I_s/eV	P^+/eV
Benzene	9.17	7.58	1.59
Naphthalene	8.12	6.4	1.72
Anthracene	7.36	5.70	1.66
Tetracene	6.89	5.10	1.79
Pentacene	6.58	4.85	1.73
Pyrene	7.37	5.8	1.57
Perylene	6.90	5.2	1.70
Chrysene	7.51	5.8	1.71
Benz[a]anthracene	7.38	5.64	1.74
Dibenz[a,h]anthracene	7.35	5.55	1.80
Coronene	7.25	5.52	1.73
p-Terphenyl	7.9	6.1	1.8
Tetrathiafulvalene (TTF)	6.4	5.0	1.4
Dimethyl TTF	6.0	5.1	0.9
Tetramethyl TTF	6.03	4.9	1.13
Dibenzo TTF	6.68	4.4	2.28
TCNQ	9.5	7.4	2.1
TNAP	8.5	6.0	2.5
p-Chloranil	9.74	8.1	1.64
p-Bromanil	9.59	7.4	2.19
p-Iodanil	8.58	5.6	2.98
Hexachlorobenzene	8.98	7.3	1.68
Hexabromobenzene	8.80	7.1	1.8
Hexaiodobenzene	7.90	5.9	2.0

extrinsic conduction, σ_0 will be smaller than for intrinsic conduction if the mobility is unchanged.

One example of the use of this approach is the semiconductivity of weak $\pi-\pi^*$ charge-transfer complexes, where the observed activation energies were consistent with intrinsic conduction as judged by application of equation 8.14 but also with extrinsic conduction involving Na^+ acceptor$^-$ impurities with donation from acceptor$^-$ to conduction levels. Determination of the pre-exponential factors together with the use of ESR to measure the concentration of paramagnetic acceptor$^-$ species led to the conclusion that these impurities were present in too low concentration to be the dominant contributors to the conduction unless unrealistically large values of carrier mobility were assumed[32].

In these measurements it is important that the crystal contacts should be ohmic and that the current–voltage characteristics are explored prior to the temperature-dependence of conduction. The processes which occur at the junctions between the electrodes (usually metal) and a molecular crystal must therefore now be explored.

When such a junction is made, charge flows between the metal and the crystal until the electrochemical potentials of the electrons (Fermi levels) in the two media are identical (figure 8.13). This transfer of charge creates an electric potential ϕ between the electrode and crystal, given by the Poisson equation

$$d^2\phi/dx^2 = Ne/\varepsilon\varepsilon_0 \tag{8.18}$$

where N is the excess charge density. The potential gradient at the junction extends into the molecular crystal rather than into the metal, since the

8.13.

Before contact   After contact

carrier density is lower in the crystal. The effective thickness x of the resulting boundary layer is determined by integrating equation 8.18:

$$x = \sqrt{(2\varepsilon\varepsilon_0\phi/Ne)}.\tag{8.19}$$

Thus electrons moving from the metal to the crystal must overcome a barrier ϕ_0 to enter the surface layer of the crystal, where

$$\phi_0 = \phi_M - EA_c\tag{8.20}$$

and a second barrier, eV, in order to reach the bulk, where

$$eV = E_F - \phi_M + EA_c,\tag{8.21}$$

and ϕ_M and ϕ_c are the work functions of metal and crystal, respectively, E_{F_M} and E_F are the Fermi levels of metal and crystal, respectively, and IP_C and EA_C are the solid-state ionisation potential and electron affinity of the crystal.

When an external field V_a is applied across the crystal, the Fermi levels of the two electrodes are displaced by an amount eV_a, modifying the right-hand side of figure 8.13 to give the situation shown in figure 8.14, where the internal barrier height (V') and thickness (x') are lower than those in the absence of an applied external field. This surface region of the crystal adjacent to the electrodes contains a reservoir of excess charge. When low external electric fields are applied, charge carriers generated within the crystal flow under the influence of the field in the direction from left to right of figure 8.14. The electric field gradient in the boundary layer is maintained by a corresponding flow of electrons from the metal to the crystal, and the contact is described as ohmic since the current density is directly proportional to the applied field, which influences only the carrier velocity, and Ohm's law is obeyed. As the applied field is increased, V' decreases and there is an increased probability of injection of excess charge into the crystal from the electrodes, leading to a change in the Fermi level and hence a

8.14.

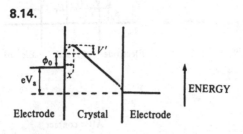

departure from Ohm's law. The excess charge is limited by the capacitance C of the crystal and its electrodes to

$$Q = CV \tag{8.22}$$

where $C = \varepsilon \varepsilon_0 / L,$ (8.23)

with L the inter-electrode spacing.

In this condition, known as space-charge-limited conduction (SCLC), the current density is given by

$$J_{SCLC} = Q/\tau \tag{8.24}$$

where $\tau = $ carrier transit time $= L^2/V\mu.$ (8.25)

Hence

$$J_{SCLC} = \varepsilon \varepsilon_0 V^2 \mu / L^3. \tag{8.26}$$

More complex derivations[33] for the case of a crystal with shallow traps give

$$J_{SCLC} = (9/8)(\varepsilon \varepsilon_0 \theta V^2 \mu)/L^3 \tag{8.27}$$

where $\theta < 1$ and reflects the extent to which traps reduce the effective mobility. The critical voltage at which the transition to such behaviour occurs may be derived simply by combining equation 8.27 with Ohm's law ($J/F = \sigma$) since the current densities for both ohmic and SCLC conduction are equal at the critical voltage, so that

$$V_t \sigma / L = (9/8)(\varepsilon \varepsilon_0 \theta V_t^2 \mu / L^3), \tag{8.28}$$

i.e. $V_t = (8/9)(L^2 \sigma / \varepsilon \varepsilon_0 \theta \mu)$
$$= (8/9)(neL^2 / \varepsilon \varepsilon_0 \theta). \tag{8.29}$$

Above the voltage V_t, the current depends on $V^2 L^{-3}$ compared with VL^{-1} for the ohmic region. As the voltage is increased above V_t the Fermi level rises in the crystal; every time it reaches the energy level of a particular set of traps, those traps become populated with charge carriers and θ is sharply reduced. In the case of a single set of traps, this would give a second critical transition voltage at which the current rose steeply from the value appropriate to the original value of θ to that for $\theta = 1$. The value of θ may thus be determined from the ratio of the current before and after the rise. This voltage, V_{TFL}, is known as the trap-filled limit voltage and is related to the trap density, N_t, by

$$N_t = \varepsilon \varepsilon_0 V_{TFL} / eL^2. \tag{8.30}$$

Furthermore, the trap depth can be deduced from the temperature-dependence of θ since

$$\theta = (N/N_t)\exp(-E_t/kT). \tag{8.31}$$

The analysis of current–voltage characteristics can thus provide a great deal of interesting information on trapping levels in molecular crystals. Figure 8.15 shows a typical current–voltage curve exhibiting the transitions mentioned above for naphthalene with silver paint electrodes[34].

A further important use of current–voltage data is that the study of temperature-dependence of conduction in both ohmic and SCLC regions can assist the interpretation of the activation energies thus derived. Roberts and Schmidlin[35] developed a general analysis of the semiconductor statistics problem in which they proposed that any linear region in a plot of log(conductivity) versus $1/T$ corresponded to a conduction process involving just two overwhelmingly dominant levels contributing to the carrier population. In the case of impurity levels, their depth and density determines whether or not they are dominant, with levels further from band edges dominating at lower densities. By definition, in SCLC-injected carrier concentrations exceed the thermal-equilibrium carrier concentrations under ohmic conditions, so that SCLC is always extrinsic. Thus, if the temperature-dependences of SCLC and ohmic conduction are identical,

8.15. Current–voltage characteristics for naphthalene.

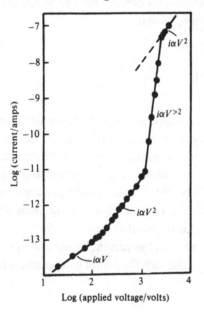

the ohmic region must be extrinsic, whereas if they are different the ohmic region must be either intrinsic or pseudo-intrinsic (alternatively described as non-extrinsic) with equal density of states for hole and electron dominant levels.

When the conductivity is not intrinsic, discrete electron-donor and/or acceptor levels from impurities or defects are present, in addition to the valence and conduction levels, influencing the carrier density. In this case the expression for carrier density as a function of temperature is derived by first determining the conditions for overall charge neutrality in the crystal, then using appropriate densities of states and energy levels with Fermi–Dirac statistics. In the general case, with both donor and acceptor levels present, electrical neutrality requires that the number of positive holes (p) in the valence band plus the number of ionised-donor levels (n_d^+) should equal the number of electrons (n) in the conduction band plus the number of ionised-acceptor levels (n_a^-):

$$p + n_d^+ = n + n_a^-. \tag{8.32}$$

A general solution to the equations which arise when these quantities are expressed in terms of densities of states and energy levels would be very complex, so certain simplifying cases corresponding to the most common conditions encountered in practice will be considered.

(1) If only donor impurities are present, charge-neutrality requires that

$$n = n_d^+, \tag{8.33}$$

i.e. $N_c \exp[(E_F - E_c)/kT] = N_d/\{1 + \exp[(E_F - E_d)/kT]\}$. (8.34)

Whence

$$n = (N_c/2)\{\exp[(E_d - E_c)/kT]\}\{-1 \pm \sqrt{(1 + [4N_d/N_c]} \\ \exp[(E_c - E_d)/kT]) \tag{8.35}$$

and for high donor concentrations where

$$[4N_d/N_c]\exp[(E_c - E_d)/kT] \gg 1,$$
$$n \approx \sqrt{(N_c N_d)}\exp[(E_d - E_c)/2kT]. \tag{8.36}$$

For lower donor concentrations and high temperatures, essentially all the donor levels are ionised so that the carrier concentration may be constant, determined by N_d, until the temperature becomes sufficient for intrinsic conduction, with higher activation energy, to dominate. Such a region of constant carrier concentration is known as an exhaustion region.

166   *Electrical properties*

(2) If both N_d donor levels and N_a acceptor levels are present, with the acceptor levels taking virtually all the charge from the n_d^+ ionised donors, the charge-neutrality condition becomes

$$n_d^+ = N_a, \tag{8.37}$$

i.e. $\quad N_d/\{1+\exp[(E_F-E_d)/kT]\} = N_a \tag{8.38}$

and hence

$$n = N_c\{(N_d/N_a)-1\}\exp[(E_d-E_c)/kT]. \tag{8.39}$$

This is known as the compensated extrinsic case, as the acceptor levels compensate the effect of the donor levels to some extent.

(3) If the donor levels lie below the Fermi level and the acceptor levels are above it but still well below the conduction band (figure 8.16), and these two levels E_q and E_m are the dominant ones, the charge-neutrality equation is

$$n_m = p_q \tag{8.40}$$

i.e. $\quad N_m\exp[(E_F-E_m)/kT] = N_q\exp[(E_q-E_F)/kT], \tag{8.41}$

whence

$$E_F = (E_m+E_q)/2 + (kT/2)\ln(N_q/N_m) \tag{8.42}$$

and

$$n = N_c\sqrt{(N_q/N_m)}\exp[(E_c-E_m)/2kT+(E_c-E_q)/2kT]. \tag{8.43}$$

This is known as the non-extrinsic or pseudo-intrinsic case.

Comparison of equations 8.17, 8.36, 8.39 and 8.43 shows that the

8.16.

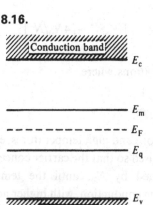

observed activation energy E in the equation $\sigma = \sigma_0 \exp(-E/kT)$ may be interpreted in a variety of ways according to the nature of the levels which dominate the conduction process, as summarised in Table 8.4. As mentioned above, if a single slope is observed in a plot of $\log \sigma$ versus $1/T$, the interpretation of this as intrinsic is possible if calculated and observed band-gaps and densities of states agree, while if the slopes are identical for ohmic and SCLC conduction it must be extrinsic. This still leaves a range of possible interpretations for the latter extrinsic case. One way in which these may be distinguished in favourable cases is to measure the conductance over as wide a temperature range as possible.

As temperature is increased, the position of the Fermi level is gradually changed and the dominant conduction process may therefore change too. A simple example mentioned in case (1) above was the transition from extrinsic conduction dominated by a set of donor levels to intrinsic conduction at high temperature, with a possible temperature-independent exhaustion region in between. Using the data in table 8.4 it is possible to predict the changes of slope of $\log \sigma$ versus $1/T$ plots for transitions between the various types of conduction process. These are summarised in table 8.5

Table 8.4 *Interpretation of experimental conduction activation energy*

Semiconductor type	Charge-neutrality condition	Activation-energy interpretation
Intrinsic	$n = p$	$(E_c - E_v)/2$
Extrinsic, single donor level	$n = n_d^+$	$(E_c - E_d)/2$
Compensated extrinsic	$n_d^+ = N_a$	$(E_c - E_d)$
Pseudo-intrinsic (non-extrinsic)	$n_m = p_q$	$(E_c - E_m) + (E_m - E_q)/2$

Table 8.5 *Interpretation of slope changes in $\log \sigma$ versus $1/T$ plots*

Transition type	Origin	Slope change (as T increases)
Non-extrinsic → non-extrinsic	Change in dominant majority-carrier level	Decrease by $(E_{m_2} - E_{m_1})/2$
	Change in dominant majority-carrier sign	Increase by $(E_q - E_v) - (E_c - E_m)$
	Change in dominant minority-carrier level	Increase by $(E_{q_1} - E_{q_2})/2$
Extrinsic → non-extrinsic		Increase by $(E_m - E_q)/2$
Non-extrinsic → extrinsic		Decrease by $(E_m - E_q)/2$

(neglecting any effects resulting from the temperature dependence of mobility). Transitions involving a change in sign of the majority carrier can be identified by the associated change in sign of the Seebeck coefficient from thermo-electric power measurements. Hence all the cases in table 8.5 can be distinguished in favourable cases.

For example, figure 8.17 shows the temperature-dependence of ohmic and SCL currents for metal-free phthalocyanine in vacuum. The slope of the low-temperature ohmic region is identical with that of the SCLC region and must therefore correspond to extrinsic conduction. At higher temperature the slope increases as expected for the extrinsic to non-extrinsic case, the increase being

$$(E_m - E_q)/2 = 1.05 - 0.38 = 0.67 \,\text{eV},$$

i.e. $E_m - E_q = 1.34 \,\text{eV}$ and $E_c - E_m = 0.38 \,\text{eV}$. Hence $E_c - E_q = 1.72 \,\text{eV}$. This interpretation attributes extrinsic conduction to donor levels located 0.38 eV below the conduction level, with a second dominant level, E_q, 1.72 eV below the conduction level at high temperature. Estimates of the density of states at E_q derived independently are consistent with this level being the top of the valence band, but it should be noted that this could not be deduced from the slope changes alone.

8.17. Temperature-dependence of ohmic and SCL currents in metal-free phthalocyanine. (After D.F. Barbe and C.R. Westgate, *Solid State Comm.*, 1969, 7, 563.)

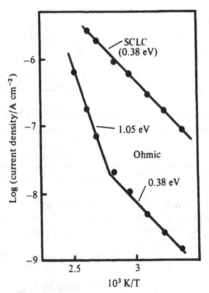

The discussion of charge-carrier generation and transport processes up to this point may be summarised by saying that the interpretation of the temperature-dependence of conduction in molecular crystals requires measurements of conductivity and mobility on highly pure and perfect crystals over wide temperature ranges in both ohmic and SCL conditions. There are very few molecular crystalline materials which can be obtained in sufficient purity and quality and which are also sufficiently thermally stable to permit such complete measurements. Thus, although such studies are of fundamental importance and considerable elegance, interest in the electrical properties of molecular crystals would not have developed to its present level without additional motivation. Such motivation stems from three important observations, namely that charge-transfer salts and related materials exhibit much higher conductivities, forming organic metals and even superconductors; that the surface conductivity of poorly-conducting molecular crystals can be enormously enhanced in the presence of certain gases, giving rise to a range of chemical sensor applications; and that the materials are photoconductors in many cases with potential applications in optical sensing, photocopying and solar photovoltaic cells.

8.6 Highly conducting organic solids

The observation of high electrical conductivity in organic solids dates back almost to the beginning of the modern development of organic semiconductor studies, to 1954 when Akamatu, Inokuchi and Matsunaga[36] reported that solid complexes between polycyclic aromatic hydrocarbons and halogens had conductivities in the range $1-10^{-3}\,\Omega^{-1}\,cm^{-1}$. It is interesting that related work, for example on halogen complexes of phthalocyanines[37], is still today a topic of strong current research activity.

Highly conducting organic solids fall into three classes according to the temperature-dependence of conductivity[38], as follows:

I. Systems whose conductivity increases strongly with temperature, with linear $\log \sigma$ versus $1/T$ plots corresponding to conduction activation energies in the range 0.3–0.5 eV, and room-temperature conductivities in the range $10^{-6}-10^{0}\,\Omega^{-1}\,cm^{-1}$ (e.g. [alkali metal]$_n$[TCNQ]$_m$).

II. Materials whose conductivity increases gradually with temperature at low temperatures, passing through a broad maximum before decreasing at high temperatures, the maximum conductivity being typically twice the room-temperature value of around $100\,\Omega^{-1}\,cm^{-1}$ (e.g. *N*-methylphenazinium TCNQ).

III. Materials showing a sharp maximum in conductivity at low temperature, with the maximum conductivity in some samples over 100 times the room-temperature value, which is commonly in the range $500–1000\,\Omega^{-1}\,cm^{-1}$ (e.g. TTF–TCNQ).

Examples of the form of the temperature-dependences for these three cases are shown in figure 8.18.

All of these highly conducting compounds share two features: they contain planar π-donor and/or acceptor molecules in segregated stacks (homosoric: see chapter 3); and they contain unpaired electrons which can move along these stacks much more easily than between them. They are thus approximately one-dimensional conductors. In addition, it has been shown that metallic conductivity persisting to low temperatures requires that

(1) the distances between neighbouring molecules within a stack must be identical;

(2) there should be some inter-stack interaction;

(3) electron–electron repulsions should be minimised.

The importance of uniform intermolecular distances within the one-dimensional conducting stacks arises simply from the band model. As in a conventional metal, the overlap of a large number of uniformly spaced units, each containing insufficient electrons to fill the available orbitals

8.18. Temperature-dependence of conductivity for some highly conducting organic solids.

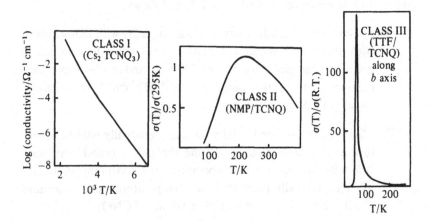

completely, leads to the formation of the part-filled bands of energy levels required for metallic conductivity.

Highly conducting organic solids differ from inorganic metals in several key respects, however. First, the overlap between adjacent units is highly anisotropic, leading to the formation of one-dimensional bands rather than the three-dimensional bands of conventional metals. Secondly, the weak intermolecular forces lead to a high probability of defects even in regular one-dimensional structures, and these severely limit the mean free path of charge carriers in otherwise ideal bands. One reason for the importance of inter-stack interactions in preserving metallic conductivity to low temperatures is that they provide pathways round defects in the one-dimensional chains. Thirdly, the molecules in an organic solid are separated by distances of the order of van der Waals contacts, so that some shortening of these intermolecular spacings is possible without such severe consequences for electron repulsions as would be encountered for species which started at separations already significantly shorter than van der Waals contacts.

It is this feature which frequently proves fatal to the preservation of uniform metallically conducting structures to low temperatures, since distortions of a uniform one-dimensional chain of species containing degenerate unpaired electrons are otherwise favoured, for they lift the degeneracy and lower the electronic energy of the systems. As shown in figure 8.19, formation of a weak dimer following the approach of two molecules, each containing one unpaired electron, leads to the formation of bonding and antibonding states, with both electrons naturally occupying the lower, bonding orbital. Extension of this to a solid composed of such dimers leads to formation of filled and empty bands, with a band-gap so long

8.19.

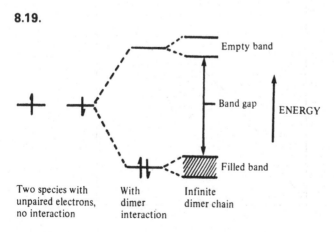

Two species with unpaired electrons, no interaction

With dimer interaction

Infinite dimer chain

as the bandwidth remains narrow (which is the case for weakly overlapping species in molecular crystals). Therefore, a distortion from a uniform stack to a stack of alternate short–long spacings lowers the electronic energy at the expense of some increase in electron repulsion energy. Such a distortion is known as a Peierls distortion[39], and has some similarity with Jahn–Teller distortions, which may be more familiar from undergraduate chemistry or physics courses. The net energy gain from such a distortion is generally less than thermal vibration energies except at low temperatures, so Peierls distortions are generally observed only at low temperatures and may be detected from a marked drop in conductivity (e.g. figure 8.18, TTF/TCNQ). The structural distortion may also be detected by X-ray diffraction, and the structural change is accompanied by a change in the phonon dispersion curve corresponding to a mode-softening (or frequency-lowering) in the wave-number region (k_F) close to the original Fermi level. This is known as a Kohn anomaly[40] and is illustrated in figure 8.20.

Associated with the lattice distortion is a degree of charge localisation, resulting in a periodic distribution of charge within the crystal, known as a charge-density wave. In the simple case in figure 8.19, where there is one unpaired electron per molecule, the resulting charge-density wave is commensurate with the lattice distortion, but this need not in general be the case for different charge densities per unit. The onset of distortions and the resulting metal : insulator transitions can be retarded by chemical modification of the system. Thus, use of large atoms (e.g. selenium instead of

8.20. Kohn anomaly.

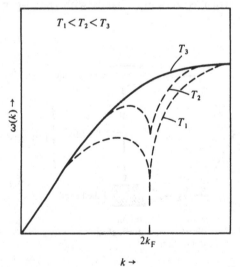

sulphur) or substituents (e.g. hexamethylenetetrathiafulvane, HMTTF) increases repulsion energies and lowers the temperature at which distortions can occur. (The Peierls transition temperatures for TTF/TCNQ, TSeF/TCNQ, HMTTF/TCNQ and HMTSeF/TCNQ are 53, 29, 48 and 24 K, respectively)[41].

An alternative approach to hindering distortions is via careful choice of the counter-ion species in the non-conducting stack. Thus, one factor aiding distortions is inter-chain coupling of the conduction electrons on one stack with the ions of the adjacent stack. If these ions are chosen to be large polarisable species, preferably disordered, any such inter-chain coupling becomes uniform and does not favour distortion. One such cation is 1,2-bis(1-ethyl-4-pyridinium)ethylene (DEPE) (figure 8.21), which forms the hydrated salt $DEPE^{2+}[TCNQ)_4]^{2-}(H_2O)_x$. Although the hydrated crystals are semiconductors with conductivity $50\,\Omega^{-1}\,cm^{-1}$, partial dehydration creates cation disorder, yielding a material of conductivity 100–$500\,\Omega^{-1}\,cm^{-1}$ which remains metallic down to 30 mK, while an anhydrous form of the material is crystallographically ordered, with a lower conductivity $(2 \times 10^{-3}\,\Omega^{-1}\,cm^{-1})^{42}$.

Requirement (3) refers both to repulsions between neighbouring molecules in a stack and to those between electrons on an individual molecule. The importance of partial charge-transfer or mixed-valence character in the stacked molecules for creating attraction rather than repulsion within homosoric stacks was discussed in chapter 3. Electron–electron repulsions within a given molecule are formally important for charge transport along the stacks[43]. Movement of charge onto a molecule already at least partially negatively charged, for example, will increase repulsion and this effect may be minimised using molecules in which any excess charges are kept apart. Thus, in TCNQ any negative charges are localised well apart on the $=C(CN)_2$ groups, whereas in TCNQF$_4$ (tetrafluorotetracyanoquinodimethane) they are moved more towards the centre of the molecule by the electronegative fluorine atoms, so that repulsion is increased.

A useful measure of the ability of a molecule to accommodate excess charge without producing excessive electron–electron repulsion is the difference between the electrochemical half-wave potentials for addition of

8.21.

the first and second electrons ($E_{\frac{1}{2}}^{-}$ and $E_{\frac{1}{2}}^{2-}$) (table 8.6). A large difference between these two half-wave potentials corresponds to a large additional electron–electron repulsion on addition of the second electron. Thus, in TCNE the small size of the molecule leads to high electron-repulsion energies and this prevents the formation of highly conducting compounds, notwithstanding the fact that this electron acceptor has a high electron affinity (2.89 eV). Similarly, the high repulsions in TCNQF$_4$ prevent it forming highly conducting complexes, whereas TCNQ and TNAP, despite their lower electron affinities, do form complexes showing metallic conductivity. It has also been suggested[44] that the creation of aromaticity when molecules such as TCNQ accept an electron is an important factor in achieving high conductivity. Delocalised aromatic π-electron systems stabilise the charged species, reducing electron–electron repulsions and increasing polarisability. Similar arguments may be applied to donor molecules which gain aromaticity following electron loss (e.g. TTF), and may be extended to charge-transport processes. Thus, charge transfer between a pair of TCNQ$^-$ ions results in a net loss of aromaticity, whereas that between TCNQ$^-$ and TCNQ^0 or between two TTF$^+$ conserves aromaticity (figure 8.22).

If the electron repulsions are minimised so that homosoric structures are formed, it is clear that the intermolecular π-electron overlap will be substantially greater between neighbouring molecules in one stack than it is between molecules in neighbouring stacks. This frequently results in a narrow-band model being applicable for charge transport along the stacks but not for charge transport between stacks (i.e. carrier mean free path along a stack $\geq$ intermolecular spacing within a stack, but carrier mean free path in direction between stacks < inter-stack spacing), and this is the strict definition of a one-dimensional conductor. The apparently diverse temperature-dependences of conductivity for classes I, II and III of organic conductors have all been fitted to a single general equation[38], which may be

Table 8.6 *Reduction potentials and electron affinities of some electron acceptors*

Acceptor	$E_{\frac{1}{2}}^{-}$/V	$E_{\frac{1}{2}}^{2-}$/V	Difference/V	EA/eV
TCNE	+ 0.15	− 0.57	0.72	2.89
TCNQ	+ 0.13	− 0.29	0.42	2.84
TCNQF$_4$	+ 0.52	+ 0.03	0.49	3.20
TNAP	+ 0.21	− 0.17	0.38	$\approx$ 2.84

interpreted in terms of narrow-band theory, of the form

$$\sigma_n(T) = AT^{-\alpha}\exp(-\Delta/T) \tag{8.44}$$

where the normalised conductivity $\sigma_n(T)$ is defined as

$$\sigma_n(T) = \sigma(T)/\sigma(295\,\mathrm{K}) \tag{8.45}$$

and A, α and Δ are constants.

In equation 8.44 the $T^{-\alpha}$ term arises from a temperature-dependent mobility while the $\exp(-\Delta/T)$ term corresponds to an activated charge-carrier concentration. In class I compounds with high activation energies, the exponential term dominates, with $\Delta \geq 2000\,\mathrm{K}$ ($\approx 0.2\,\mathrm{eV}$) and $\alpha \geq 4$. At the other extreme, class III compounds have $\Delta = 0$ and the rapid increase in conductivity on cooling corresponds to a dominant $T^{-\alpha}$ term with $\alpha \geq 4$.

8.22.

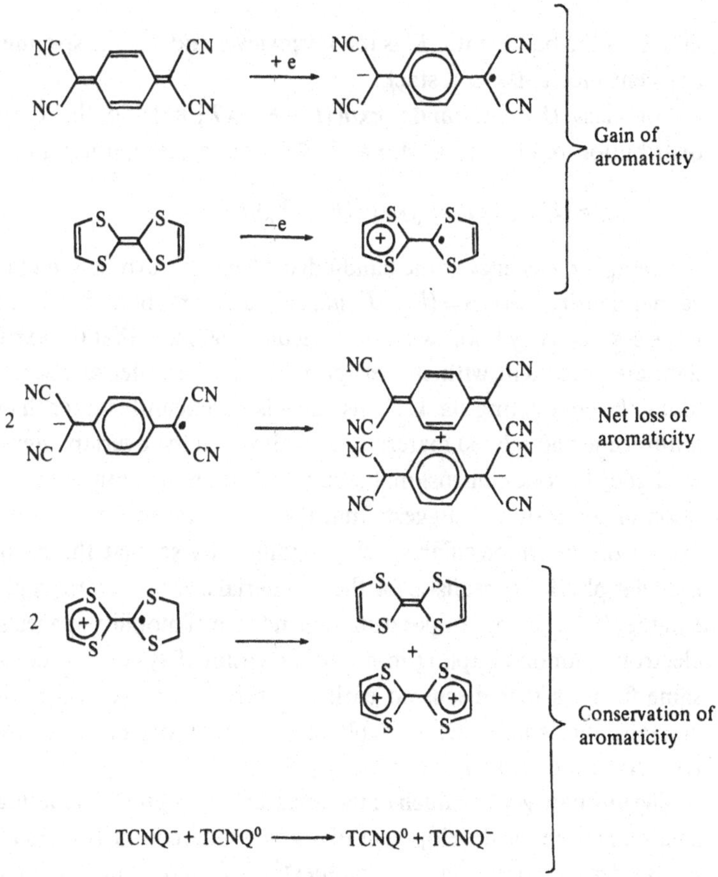

Gain of aromaticity

Net loss of aromaticity

Conservation of aromaticity

TCNQ⁻ + TCNQ⁰ ⟶ TCNQ⁰ + TCNQ⁻

(The drop in conductivity at very low temperatures for class III compounds is a consequence of the Peierls distortion discussed earlier and is not covered by equation 8.44.) For class II compounds, both terms contribute, with Δ in the range 300–900 K typically, so that the conductivity passes through a broad maximum. The constant A is chosen so that $\sigma_n(295\,\text{K}) = 1$ as required by equation 8.45. For this class II general case, a one-dimensional narrow-band model may be used to derive the carrier concentration, η, per unit length:

$$\eta = \int_0^{E_0} [1 + \exp\{(E - E_F)/kT\}]^{-1} \rho(E)\mathrm{d}E \qquad (8.46)$$

where $\rho(E)$ is the number of states per unit energy range:

$$\rho(E) = (2/\pi a)\{1/\sqrt{[E(E_0 - E)]}\} \qquad (8.47)$$

and E_0 is the bandwidth, E_f is the Fermi level and a is the spacing between adjacent molecules in a stack.

For class II compounds, $\exp[(E - F_F)/kT] \gg 1$, so the Fermi–Dirac distribution reduces to a Maxwell–Boltzmann distribution and

$$\eta \approx (2/a)\sqrt{(kT/\pi E_0)}\exp[(E_F - E_b)/kT], \qquad (8.48)$$

where E_b is the energy of the band-edge. Using this expression to derive the carrier density, with $\Delta = (E_b - E_F)/k$, the carrier mobility can be calculated if the conductivity is known. Such calculations show that the experimental data are consistent with carrier mobilities of the order of $10\,\text{cm}^2\,\text{V}^{-1}\,\text{s}^{-1}$ along the conducting stacks. This value is significantly larger than that for anthracene and related materials, and although the overlap integral J may be higher for one-dimensional conductors than for anthracene, consideration of equation 8.4 suggests that the carrier mean free path is likely to exceed the intermolecular spacing significantly so that the narrow-band model is physically realistic for these materials. However, the high value of α in the $T^{-\alpha}$ term for temperature-dependence of mobility suggests that the electron–phonon coupling in a one-dimensional system cannot be of the same form as that discussed earlier in this chapter for other molecular crystals, and is a further example of the complexity of the whole area of temperature-dependence of mobility.

The ultimate goal of much of the research into highly conducting organic solids has been the development of organic superconductors. In the theory of Bardeen, Cooper and Schrieffer[45], superconductivity arises as a consequence of highly organised motion of pairs of conduction electrons with equal and opposite momentum and spin. The motion of one electron disturbs the surrounding lattice in a way which causes a deflection of the second electron towards it, thus creating an effective attraction force which reduces the scattering from irregularities in the solid lattice and produces superconductivity. In 1964 Little[46] suggested that similar attractive forces might be produced much more readily by disturbances of polarisable *electron* clouds (rather than of the much heavier *nuclei*) accompanying motion of the first electron, and that these might therefore permit superconductivity at much higher temperatures. In 1979, 25 years after the first observation of highly conducting organic solids, superconductivity was finally observed for the first time in an organic solid [(tetramethyltetraselenafulvalene)$_2$PF$_6$] near 1 K under 12 kbar pressure[47]. More recently, ambient-pressure organic superconductors have been developed, such as (TMTSF)$_2$ClO$_4$ and the bis(ethylenedithio)tetrathiafulvalene (BEDT-TTF) salts with various counter-ions (e.g. the AuI$_2^-$ salt, which remains superconducting to about 5 K[48]). Figure 8.23 shows the crystal structure and low-temperature electrical conductivity of one of these materials (see also chapter 9).

It is not yet possible to predict the likelihood that these exciting materials are the forerunners of a series of compounds with successively higher superconducting transition temperatures, possibly leading to practical high-temperature superconductors. However, it is already clear that highly

8.23.

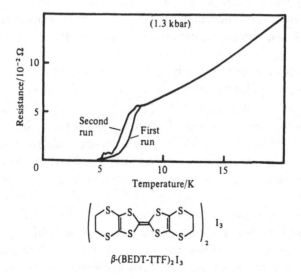

β-(BEDT-TTF)$_2$I$_3$

conducting organic solids in general share characteristics which will limit their practical application to specialised devices. Their anisotropic conductivity is susceptible to defects and impurities and implies that their true electrical properties can be effectively studied and exploited only with single-crystal samples. Furthermore, their tendency to crystallise in several phases, not all of which are highly conducting, complicates production methods for any devices using such materials. Finally, although the possibilities for modifying organic solids to improve electrical properties are very wide, present indications are that many highly conducting materials involve expensive and difficult-to-handle compounds (e.g. TTF, TSeF, TTeF and analogues) whose chemical and physical stability is inferior to inorganic electronic materials.

8.7   Effects of gases on electrical conductivity of molecular crystals

One application of organic semiconductors in which these material limitations are less severe is chemical sensing[49]. Since the earliest studies of electrical conductivity in organic solids, it has been realised that electron-accepting gases may influence conductivity via surface (and possibly also bulk) charge-transfer interactions which facilitate generation of charge-carrying positive holes. If the majority carrier in vacuum conditions is already positive, exposure to electron-accepting gases will increase the carrier concentration and hence also the conductivity. Conversely, if the majority carrier is initially negative, the positive holes produced by electron-accepting gases will recombine with the original carriers, reducing the conductivity in a depletive chemisorption process. Analogous effects, with carrier signs reversed, are expected on exposure to electron-donating gases.

These effects were used originally to determine the sign of the majority charge carriers in molecular crystals of different types[50]. More recently[51] they have been studied in a more quantitative manner, and surface conductivity changes by factors of up to 10^8 have been observed on exposure to pressures of only 10^3 Pa of electron-accepting gases such as NO_2. The use of single-crystal samples with good-quality surfaces is important for the clear interpretation of such observations. Experiments on samples with an earthed guard-ring (figure 8.24) show that in most cases the conduction enhancement is confined to the surface layers, vanishing when the guard-ring is earthed. For compressed pellets and films, diffusion along internal surfaces can complicate the discrimination of surface and bulk effects and it becomes important to distinguish effects caused by adsorbed vapours reducing inter-particle resistance from genuine electronic effects on the materials of the particles themselves. Thus, for example, experiments on the effects of vapours on the electrical properties of β-carotene extracted from steer's noses, which were intended to investigate possible mechanisms of the sense of smell, showed conductivity increases of 10^6 when compressed pellets were exposed to methanol vapour[52] yet no increases when single crystals were similarly exposed[53].

Figure 8.25 shows the conductivity of a sublimed film of lead phthalocyanine at 150 °C as a function of the concentration of NO_2 in dry air surrounding the film[54]. If the conductivity increase is assumed to depend on the fractional coverage θ of the surface by NO_2, the observed linearity of the log(current) versus log(NO_2 concentration) plot corresponds to the Freundlich adsorption isotherm $\theta = kP^x$. The conductivity increases reverse on removal of the NO_2, consistent with weak chemisorption involving charge transfer rather than chemical reaction. At high concentrations of NO_2 the effect saturates as the surface coverage

8.24.

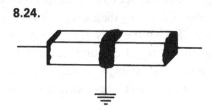

8.25. Conductivity of a lead phthalocyanine film at 150 °C as a function of NO_2 concentration in air.

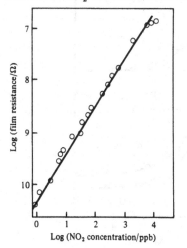

approaches unity, and the limiting surface resistivity at room temperature is in the range 10^6–$10^8\,\Omega$ square $^{-1}$, depending only slightly on the nature of the metal in the phthalocyanine complex. A surface resistivity of this order is equivalent to a bulk resistivity of approximately $1\,\Omega$ cm if the surface layer is assumed to be $10\,\text{Å}$ thick, and it is interesting to compare this with the value of about $2 \times 10^{-3}\,\Omega$ cm reported for single crystals of $NiPcI_{1.0}$ in which nickel phthalocyanine is uniformly doped throughout the bulk with the electron-acceptor iodine[55]. In view of the likely greater irregularity of crystal surface compared with bulk structure, these figures suggest that the adsorbed gas effectively creates a surface layer of almost-metallic conductivity, as expected for partial oxidation of the homosoric stacks of molecules in phthalocyanine crystals.

The effects of gases on the magnitude, rate and reversibility of conductivity changes have been studied for a wide range of molecular crystal materials including aromatic hydrocarbons, metal complexes of macrocyclic ligands, electron acceptors (e.g. TCNQ), dyes and charge-transfer complexes. The strength of the surface charge-transfer interactions is the single most important factor controlling these properties. Thus, if the interactions are too weak their effect on the activation energy for carrier generation is small, whereas if they are too strong and localised the charge carriers are coulombically bound to the adsorption sites and conduction enhancement is small. Figure 8.26 shows these effects for too-weak

8.26.

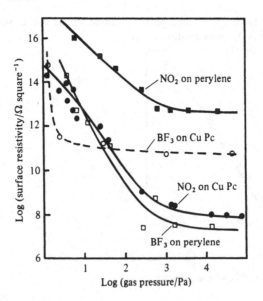

adsorption (NO_2 on perylene), too-strong adsorption (BF_3 on phthalocyanines) and optimum adsorption (NO_2 on phthalocyanines, BF_3 on perylene). Both the rate of the conductivity change on exposure to the gas and the rate of its reversal in clean conditions are optimised when the surface charge-transfer interactions are as weak as possible consistent with the retention of conductivity effects of reasonable magnitude. While it is obvious that the reversal will be faster for more weakly bound gases, the faster rise in conditions of weaker surface charge-transfer is less obvious, until it is appreciated that adsorption of a gas is primarily limited kinetically by the ease of displacement of adsorbed oxygen from the surface. Under all normal atmospheric conditions, organic solid surfaces may be assumed to be completely covered with an adsorbed layer comprising mainly oxygen species since, for example, it has been shown that a monolayer of oxygen forms on a clean phthalocyanine surface in less than one minute on exposure to a pressure of only 10^{-4} Pa of oxygen. Weaker surface charge-transfer interactions weaken the binding of the oxygen to the surface and thus enhance the rate of its displacement by stronger electron acceptors[56].

These effects of gases on electrical conductivity of molecular crystals, particularly phthalocyanines, provide the basis of a sensitive device for detecting electron-accepting gases in the atmosphere[54]. The principal problems in using such a device are the complex response and reversal kinetics, and the effects of humidity. Response and reversal kinetics follow the Elovich equation

$$d\theta/dt = a\exp(-b\theta), \tag{8.49}$$

where a and b are constants, so the surface coverage, θ, (and hence conductivity) varies linearly with log(time) rather than directly with time. Although responses to small concentrations of electron-accepting gases in nitrogen carrier gas or admitted to the material in a vacuum chamber are rapid, those in air occur over several hours for simple planar phthalocyanines, even at temperatures up to 200 °C. Furthermore, freshly prepared films initially show extremely small responses. The development of faster and larger responses following prolonged exposure has been explained by a kinetic model in which large surface dipoles (at sites where residual electron-acceptor species are strongly bound) repel neighbouring oxygen molecules on the surface of the phthalocyanine, speeding up their desorption[57]. This model has led to three approaches to obtaining organic semiconductor sensors able to give rapid indications of gas concentrations.

In the first of these, reproducible response kinetics are achieved by commencing sampling pulses only when the background conductivity has reached a standard reference value which reflects a standard residual surface coverage by the electron acceptor species. Under these conditions, the response in a short time is reproducibly related to the final equilibrium response. However, this approach requires microprocessor control and complex sampling pumps and valves[58]. The second approach seeks to destroy the strongest adsorption sites by chemical reaction, by heating the phthalocyanine to 360 °C in air. This only works for the more reactive non-planar lead phthalocyanine and provides films which respond rapidly at room temperature. However, they still show some slow components of response resulting from the incomplete reaction of the strong sites, and the response is severely inhibited by humidity[59]. The final approach is to minimise the number of strong adsorption sites which lead to the problem of slow response. Figure 8.27 shows a schematic of the heterogeneous surface of a phthalocyanine film. This illustrates the fact that charged species at surface projections benefit less from polarising the surrounding material than do charged species adsorbed in crevices on the film surface. From the energetics of charge transfer described in figure 8.12 and equation 8.14 it is clear that the higher the polarisation energy, the easier is the charge transfer and therefore the stronger the adsorption. Therefore, to avoid strong adsorption it is necessary to use surfaces which are as uniform as possible and to attempt to reduce the local polarisability. Both objectives can be achieved by using the self-assembly principles discussed in section 3.5, particularly the use of crown-ether-substituted phthalocyanines[60] and discotic liquid crystalline phthalocyanines[61]. As previously explained, the

8.27. The heterogeneous surface of a phthalocyanine film. Polarisation energy at site A > B > C > D > E > F.

self-assembly ensures uniform films, while the polarisation energy (shown in section 8.2 to vary as r^{-4}, where r is the distance from the charged species) is reduced by the presence of the less polarisable σ-bonded substituents. These materials give rapid reversible responses to electron-acceptor gases at low temperatures. However, the problem of interference by water vapour remains serious, particularly in the case of the crown-ether-substituted materials, where the crown-ether oxygens provide many opportunities for hydrogen bonding to water.

8.8   Photoconductivity

For many molecular crystals the energy required for intrinsic charge-carrier generation is much closer to the typical energies at which optical absorption takes place in the crystals (1–4 eV) than to thermal energy (0.026 eV at room temperature). Thus even poorly conducting molecular crystals exhibit photoconduction, and, as pointed out at the beginning of this chapter, it was the observation of photoconductivity in anthracene which began the exploration of electrical properties of these materials. The many subsequent studies of photoconductivity have shown that this provides many more accessible experimental parameters than do studies of semiconductivity. They have also provided an understanding of the mechanisms of photoconduction and hence led to a range of practical applications of organic photoconductors. These include photocopying[62], solar photovoltaic cells[63] and photo-electrochemistry[64].

The formation of Frenkel, charge-transfer and Wannier excitons in molecular crystals following absorption of light has been discussed in chapter 6. Charge-transfer and Wannier excitons involve positive and negative charges on different molecules with a charge separation such that coulomb forces of attraction are still significant. Thermal (or field-assisted) dissociation of these 'geminate pairs' produces charge carriers separated by larger distances, at which the local polarisation energies of the charges become comparable to or greater than the residual coulomb attraction between the carriers, so that they no longer spontaneously drift together to recombine. In the case of Frenkel excitons, the mechanisms for formation of the initial geminate pairs must be less direct, as the initial photon absorbed has sufficient energy only to excite electrons to higher orbitals on the same molecule. These mechanisms include:

(1) migration of the exciton to an impurity or defect site or to the crystal surface, followed by exciton dissociation producing one trapped charge and one separated charge carrier;

(2) migration of the exciton to an electrode, producing photo-injection of charge carriers into the crystal;
(3) exciton–exciton collisions followed by energy redistribution, yielding a ground-state molecule and a separated carrier pair;
(4) photo-ionisation of an exciton, followed by a similar energy redistribution.

The excitons involved may be singlet states, with high probability of formation but short lifetimes, or triplet states, with low generation probability but long lifetimes.

Mechanisms (1) and (2) involve single photons and are independent of the density of excited states in the crystal. Mechanisms (3) and (4), however, involve two photons and do depend on excited-state density. The use of polarised exciting light can help to distinguish between these two groups. Since most absorption bands of crystals are polarised with respect to the crystal axes, changing the polarisation of the incident light will alter the excited-state density and the penetration depth of light into the crystal. If this has no effect on the magnitude of the photocurrent (as is the case for many crystalline charge-transfer complexes[65]), single-photon generation is likely. (However, it does not follow that if there is such an effect the process cannot be a single-photon one, since a surface exciton–dissociation process involving a single photon would also be more favoured if the light is strongly absorbed.) Similarly, carrier loss processes may be unimolecular (loss to electrodes or traps) or bimolecular (recombination). The overall dependence m of steady-state photocurrent i on light intensity L may thus be deduced, as shown in table 8.7 (where $i \propto L^m$). Additional complications occur if excitons are produced by two-photon absorption of high-intensity light (when correspodingly higher values of m occur) and if carrier mobility is reduced in the presence of high carrier concentrations (when power laws lower than 0.5 may occur).

The thermal field-assisted dissociation of geminate pairs of charges was

Table 8.7 *Dependence of photocurrent on light intensity*

Dominant generation process	Dominant loss process	Intensity dependence
Single-photon	Unimolecular	1
	Bimolecular	0.5
Two-photon	Unimolecular	2
	Bimolecular	1

discussed by Chance and Braun[66] in terms of a model developed by Onsager[67] much earlier to describe the Brownian motion of one charged particle in the coulomb potential of another in an external applied field, in relation to the variation of currents in ionisation chambers with applied field. In this model, the sphere within which two charges will spontaneously recombine is defined by the Onsager radius (or thermalisation length) r_c, where

$$r_c = e^2/4\pi\varepsilon\varepsilon_0 kT. \tag{8.50}$$

The dependence of photoconduction quantum efficiency Φ on applied electric field F, for low fields, is found using Onsager's theory to be

$$\Phi = A[1 + (er_c/2kT)F]. \tag{8.51}$$

This theory has been tested experimentally using the techniques described earlier for pulsed-light measurements of carrier-drift mobility (figure 8.2) and measuring the total area of the current pulse produced by each light pulse of known intensity and duration. The quantum yields thus determined obeyed equation 8.51 and the experimental values of the slope:intercept ratios for holes and electrons in pure anthracene were found[66] to be $3.02(\pm 0.08) \times 10^{-5}$ and $3.2(\pm 0.12) \times 10^{-5}$ cm V^{-1}, respectively, in good agreement with $(e^3/8\pi\varepsilon\varepsilon_0 k^2 T^2) = 3.38 \times 10^{-5}$ cm V^{-1} for anthracene.

A more generalised Onsager model has been developed by Noolandi and Hong[68] for geminate pairs produced by reversible auto-ionisation of Frenkel excitons of finite lifetime equal to that of fluorescence (figure 8.28). This model, in which both k_1 and k_2 are field- and direction-dependent, has been used to account for the electric field-dependence of both photocurrent and fluorescence quenching[69].

It is possible to distinguish between surface and bulk charge-carrier generation processes by comparing the photoconduction action spectrum with the absorption spectrum of the sample. If they are similar (symbiotic

8.28.

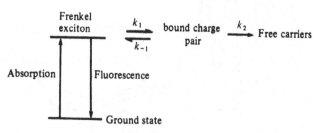

behaviour) it is likely that surface mechanisms predominate. In regions of high absorption coefficient, light is absorbed close to the sample surface favouring surface generation processes. If they are complementary, with high absorption corresponding to low photocurrent (antibatic), bulk photoconduction mechanisms are indicated, with the more weakly absorbed light penetrating further into the bulk of the sample where the preferred carrier-generation process occurs.

Figure 8.29 shows an example of symbiotic behaviour[70] for anthracene. The photocurrent in such a case can be modelled by first setting up a steady-state equation for the generation and loss of excitons for a given volume element (dx) of the sample, and then applying boundary conditions – for example, by assuming that excitons reaching the sample surface are very rapidly dissociated to charge carriers. In the steady state, the rate of exciton generation (proportional to $I_0 \exp(-\alpha x)$, where α is the absorption coefficient and I_0 the incident photon flux) equals the rate at which excitons are lost by natural decay ($-n(x)/\tau$, where τ is the exciton lifetime) and diffusion ($-D d^2 n(x)/dx^2$, where D is the exciton diffusion coefficient). The rate of carrier generation is determined by the rate of arrival of excitons at the surface by diffusion, and for thick samples where both the absorption depth of the light $(1/\alpha)$ and the exciton diffusion length 1 $(=\sqrt{(zDt)}$, see chapter 6) are much less than the sample thickness, it has been shown that the resulting expression relating photoconduction quantum yield Φ and absorption coefficient α is

$$\Phi^{-1} = (\alpha l b)^{-1} + b^{-1} \tag{8.52}$$

8.29. Photoconduction and absorption spectra of anthracene.

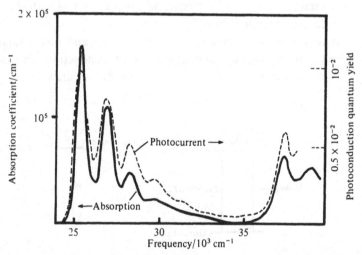

where b is a constant. Figure 8.30 shows a plot of Φ^{-1} versus α^{-1} for pure anthracene and anthracene heavily doped $(2 \times 10^{-4} \, \text{M})$ with tetracene. The intercept: slope ratio yields the exciton diffusion length l, which as expected is shorter in the heavily doped sample.

In other cases, transitions from symbiotic to antibatic behaviour have been observed as samples initial *in vacuo* are exposed to gases which adsorb and create sites for enhanced surface exciton dissociation. Figure 8.31 shows such behaviour for lead phthalocyanine, with bulk-carrier generation dominant *in vacuo* and surface generation in air[71].

A different class of gas effects on photoconductivity is exemplified by the observation of enhanced near infra-red photocurrents in phthalocyanines in the presence of oxygen[72]. In this case it is the effect of paramagnetic oxygen increasing the transition probability for singlet–triplet excitation which is responsible for the increased photocurrent, with long-lived triplet states providing effective carrier-generation pathways.

The above examples all involve Frenkel excitons. For charge-transfer complexes, intermolecular charge-transfer transitions lead more directly to charge carriers. Figure 8.32 is an energy-level diagram for charge-carrier formation and charge-transfer absorption in such a complex[65], from which it can be seen that the energy difference between the exciton and conduction levels, though small, is influenced by many factors. As expected from

8.30. Relationship between photoconduction quantum yield (ϕ) and optical absorption coefficient (α).

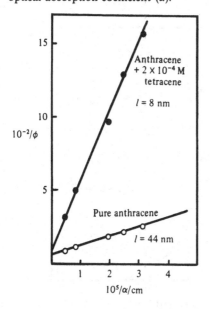

figure 8.32, photoconduction in charge-transfer complexes generally either follows the charge-transfer spectral response or peaks at the high-energy side of the charge-transfer band. However, the probability of the exciton state producing charge carriers is determined not by the energy gap (Δ) between the exciton and conduction states in figure 8.32, but rather by the kinetics of the processes involved in carrier separation, as well as by the kinetics of the competing processes such as decay to the original ground state.

A major factor influencing these kinetics is the relative orientation and overlap of adjacent molecules in the crystalline complex, and correlation of the magnitude of photoconduction with crystal structure has led to two structural conditions favouring efficient photoconduction[65]. Firstly, efficient separation of charges from initial ion-pair states in a heterosoric structure is favoured if donor *or* acceptor molecules (*but not both*) in adjacent stacks overlap. This can be seen from figure 8.33, where it is clear that such inter-stack motion (1) is favoured relative to transfer of positive charge from a donor molecule to an acceptor (2), with the only alternative being recombination via reverse of the original charge-transfer transition (3). Secondly, the rate of the recombination electron-transfer step (3) will be reduced relative to charge-carrier production processes if the overlap of the relevant orbitals (donor HOMO and acceptor LUMO generally) is poor

8.31. Photoconduction and absorption spectra of lead phthalocyanine crystal.

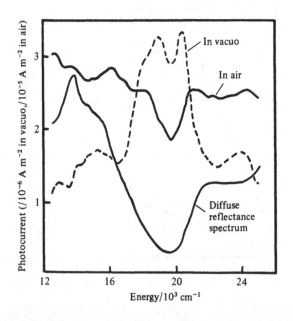

for the observed relative orientation of donor and acceptor within the stack.

Finally, it is important to realise that illumination of a sample will inevitably lead to some heating of the surface caused by radiation-less decay processes. This heating will always produce some change in the semiconduction of the sample, and great care must be taken to ensure that apparent

8.32. Energy level diagram for photoconduction in crystalline molecular complexes. (ΔW_f = energy of formation of complex DA in solid phase. A_1 and A_2 are attractive forces, other than charge-transfer resonance forces, in ground- and excited-state species, respectively. R_1 and R_2 are the corresponding repulsive forces. E is the exchange energy in excited state. R_{es_1} and R_{es_2} are perturbations of ground and excited states resulting from charge transfer resonance. P_{DA}, $P_{D^+A^+}$ and P_{A^-} are the interaction energies between DA, D^+A^-, D^+ and A^-, respectively, and the crystal lattice.)

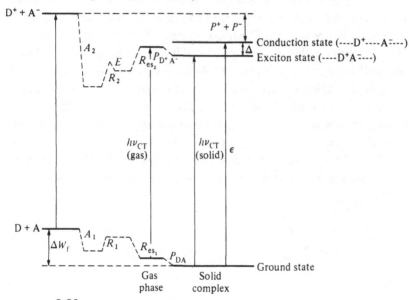

8.33.

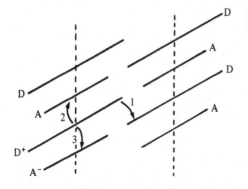

photocurrents, particularly small ones, are not artefacts of this heating effect. For example, Eley and co-workers showed that several reports of photocurrents in biological materials of high resistivity and conduction activation energy could be re-interpreted in terms of changes in the dark currents produced by heating effects of entirely plausible magnitude[72]. Heating effects are generally much slower than true photocurrent effects and may therefore be avoided if pulsed-light or chopped-light[73] techniques are used.

8.9   Molecular materials in electrophotography, electroluminescence and solar cells

Photocopiers and laser printers both work on the same principle and frequently use molecular crystalline materials for improved performance and lower toxicity and cost than corresponding inorganic alternative materials. The principle is illustrated in figure 8.34. A metal-coated base layer supports a thin organic photoconductor charge-generating layer, covered by an optically transparent hole-conducting charge-transport layer. Initially the whole multi-layer sheet or drum assembly is electrically charged by a corona discharge, the upper surface being negative. Next, the sheet or drum is illuminated with the image of the material to be copied or printed. Where the charge-generating layer is illuminated, charge carriers are produced. This process is very effective as the layer is subjected to the large electric field of the charged system, so that field-assisted charge separation

8.34.

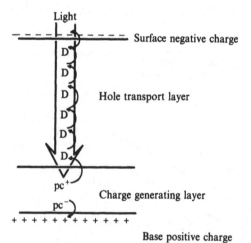

Light

Surface negative charge

Hole transport layer

Charge generating layer

Base positive charge

occurs (as predicted, for example, by the Onsager theory of equation 8.51). The negative charges immediately recombine with the positive charges on the metal base layer. However, the holes migrate through the charge-transport layer and discharge the surface negative charge in the illuminated regions. At this stage, therefore, the surface is negatively charged only in regions where the image to be copied or printed was dark. Charged toner particles, consisting of a thermoplastic resin base impregnated with carbon black and materials known as charge control agents (which determine the sign and magnitude of the charge acquired by the toner), are then attracted to the remaining negatively charged surface regions to form the image, which is transferred and thermally fused to the paper.

The ideas discussed in section 8.8 all influence the choice of organic material for the charge-generating layer. Spectral response can be tailored in organic materials, so that white-light photocopiers use organic materials such as perylene imides which give a broad photoconductive response across the whole visible region, whereas laser printers with laser diodes emitting at 820 nm use a range of phthalocyanines including the X-form of metal-free phthalocyanine (see figure 6.5, p. 106). Other advantages of organic materials for this application include easy device fabrication and flexibility of the material (so that it can be used in belt form for high-speed copiers). Photochemical degradation ultimately limits the lifetime of these materials in photocopiers and printers, leading to poorer discharge of the charge in the illuminated areas and hence the well-known dark-background effect seen in copies produced by machines with drums nearing the end of their useful life. The charge-trasport layers are commonly triphenylmethanes, hydrazones or oxadiazoles, and, although the hole mobilities in such materials are not large, the high electric fields and thin layers used mean that hole transit times are sufficiently short for rapid copying. Clearly, such organic devices are susceptible to mechanical damage, and it is therefore important to exercise extreme care when clearing paper jams etc. The importance of organic materials in these applications is reflected in the fact that there are now well over 3000 patents for organic photoconductor materials, with over 600 of these appearing in one year (1983) for example[74,75].

Organic photoconductor materials have also been used in solar photo-electric cells[75], where the prospect of using organic materials for cheap, large-area, flexible, lightweight devices in which optimal matching of light absorption to the solar spectrum could be achieved using molecular design has been the principal motivation. As in all photovoltaic devices, the principle involves photogeneration of charge carriers in one or more photoconductive layers followed by selective migration of charges to dissimilar electrodes or across junctions between n- and p-type semiconductor material. Three types of cell have been explored: Schottky cells in which a single p-type photoconductive layer is sandwiched between indium tin oxide (ITO) transparent conducting glass and metal (e.g. Al) electrodes; p/n heterojunction cells, with a junction between p- and n-type dye layers sandwiched between ITO and gold electrodes; and *pin* cells, in which the p/n heterojunction is replaced by a third region comprising a mixed dye layer. The last two types offer greater possibilities for matching cell absorption characteristics to the solar spectrum since two different dyes are used. Such cells provide open-circuit voltages up to about 0.8 V, but their efficiency is limited by the low short-circuit currents which can be sustained (maximum about $2.5\,\text{mA cm}^{-2}$). This low value is limited by the low charge-carrier mobility at ambient temperatures (c.f. section 8.2) and the short singlet exciton diffusion lengths (c.f. section 6.7), which iimit the rate at which charge carriers can be produced and transferred across the junction. These limitations cannot be avoided by use of very thin dye layers since the amount of sunlight actually absorbed would then be very small. Other practical problems in developing organic solar cells include finding methods for: (i) design of dyes with broad absorption; high absorbance; high chemical, photochemical and physical stability and suitable electrochemical properties; (ii) preparation of coherent well-structured pinhole-free dye films; (iii) effective doping of the dyes to produce clear p- and n-type behaviour; and (iv) fabrication of well-defined defect-free junctions between dye layers. Many hundreds of potential organic solar cells have been investigated, but the most promising so far are based on combinations of p-type phthalocyanines with n-type perylene biscarboximides, with efficiencies of approximately 2%. In addition to the scientific problems mentioned above, further development of these devices is limited by the economic factors which determine energy policy, and at the time of writing it seems unlikely that major development programmes will be reactivated until the costs of conventional energy sources rise substantially.

The reverse process of solar cells, namely the conversion of electric current to light via electroluminescence, has also been explored using organic materials[75]. Thus, for example, emission of electrons from a low-work-function metal (e.g. MgAl alloys) into the lowest vacant orbital of a dye, followed by radiative recombination with holes in the highest occupied orbital created by hole transfer from a hole-transport layer (e.g. a triphenylamine derivative) leads to light emission. Typically, Al–8-hydroxyquinoline complexes (green), peryleneimides (red) and styryl dyes (e.g. blue) have been used, together with a range of fluorescent dyes to modify the emission characteristics. These are potentially useful for large area colour displays with low power requirements and wide available colour range. However, efficiency and stability still present serious obstacles to practical deployment of such devices.

References

1 A. Pocchetino, *Acad. Lincei. Rendiconti*, 1906, **15**, 355.
2 D.D. Eley, *Nature*, 1948, **162**, 819.
3 A.T. Vartanyan, *Zh. Fiz. Khim.*, 1948, **22**, 769.
4 H. Akamatu and H. Inokuchi, *J. Chem. Phys.*, 1950, **18**, 810.
5 O.H. Le Blanc, *J. Chem. Phys.*, 1960, **33**, 626; R.G. Kepler, *Phys. Rev.*, 1960, **119**, 1226; W.E. Spear, *J. Non-cryst. Solids*, 1969, **1**, 197.
6 R.H. Young, E.I.P. Walker and A.P. Marchetti, *J. Chem. Phys.*, 1979, **70**, 443.
7 See, for example, J.S. Blakemore, *Solid State Physics*, Philadelphia: Saunders, 1970.
8 See, for example, A.R. West, *Solid State Chemistry and its Applications*, Chichester: Wiley, 1984.
9 J.M. Ziman, *Principles of the Theory of Solids*, Cambridge: Cambridge University Press, 1964.
10 L.B. Schein, *Phys. Rev. B*, 1977, **15**, 1024.
11 G.G. Roberts, N. Apsley and R.W. Munn, *Phys. Reports*, 1980, **60**, 59.
12 Z. Burshtein and D.F. Williams, *Phys. Rev. B*, 1977, **15**, 5769.
13 L.B. Schein, W. Warta, A.R. McGhie and N. Karl, *Chem. Phys. Lett.*, 1980, **75**, 267.
14 L.B. Schein, C.B. Duke and A.R. McGhie, *Phys. Rev. Lett.*, 1979, **40**, 197.
15 J. Berrehar and M. Schott, *Mol. Cryst. Liq. Cryst.*, 1978, **46**, 223.
16 L.B. Schein and A.R. McGhie, *Phys. Rev. B*, 1979, **20**, 1631.
17 W. Warta and N. Karl, *Phys. Rev. B*, 1985, **32**, 1172.
18 K.H. Probst and N. Karl, *Phys. Stat. Sol. a*, 1975, **27**, 499.
19 See reference 11, p. 131.
20 J. Yamashita and T. Kurosawa, *J. Phys. Chem. Solids*, 1958, **5**, 34.
21 I.G. Austin and N.F. Mott, *Adv. Phys.*, 1969, **18**, 41.
22 R.J.F. Dalrymple and W.E. Spear, *J. Phys. Chem. Solids*, 1972, **33**, 1071.
23 H. Sumi, *Solid State Comm.*, 1978, **28**, 309; *J. Chem. Phys.*, 1979, **70**, 3775.
24 H. Sumi, *Solid State Comm.*, 1979, **29**, 495.
25 L.B. Schein and A.R. McGhie, *Chem. Phys. Lett.*, 1979, **62**, 356.
26 D.D. Eley and D.I. Spivey, *Trans. Faraday Soc.*, 1960, **56**, 1432, and references therein.
27 H. Meyer and H. Neldel, *Z. Tech. Phys.*, 1937, **18**, 588, and reference 11, p. 90.
28 L.E. Lyons, *J. Chem. Soc.*, 1957, 5001.
29 P.J. Bounds and R.W. Munn, *Chem. Phys.*, 1979, **44**, 103; 1981, **59**, 41, 47.
30 I. Eisenstein, R.W. Munn and P.J. Bounds, *Chem. Phys.*, 1983, **74**, 307. I. Eisenstein and R.W. Munn, *Chem. Phys.*, 1983, **77**, 47; 1983, **79**, 189.
31 N. Sato, K. Seki and H. Inokuchi, *J. Chem. Soc. Faraday II*, 1981, **77**, 1621.
32 P.J. Munnoch and J.D. Wright, *J. Chem. Soc. Faraday I*, 1976, **72**, 1981.
33 J.S. Bonham and D.H. Jarvis, *Aust. J. Chem.*, 1977, **30**, 705.
34 M. Campos, *Mol. Cryst. Liq. Cryst.*, 1972, **18**, 105.
35 G.G. Roberts and F.W. Schmidlin, *Phys. Rev.*, 1968, **180**, 785.
36 H. Akamatu, H. Inokuchi and Y. Matsunaga, *Nature*, 1954, **173**, 168.
37 T. Inabe, T.J. Marks, R.L. Burton, J.W. Lyding, W.J. McCarthy, C.N. Kannewurf,

G.M. Reisner and F.H. Herbstein, *Solid State Comm.*, 1985, **54**, 501.

38 A.J. Epstein, E.M. Conwell and J.S. Miller, *Ann. N.Y. Acad. Sci.*, 1978, **313**, 183.

39 R.E. Peierls, *Quantum Theory of Solids*, Oxford: Oxford University Press, 1955.

40 W. Kohn, *Phys. Rev. Lett.*, 1959, **2**, 393.

41 S. Kagoshima, in *Extended Linear Chain Compounds*, Vol. 2, Ch. 7, ed. J.S. Miller, New York: Plenum, 1982.

42 G.J. Ashwell, *Nature*, 1981, **290**, 686.

43 A.F. Garito and A.J. Heeger, *Acc. Chem. Res.*, 1974, **7**, 232.

44 J.H. Perlstein, *Angew. Chem. (Int. Edn)*, 1977, **16**, 519.

45 J. Bardeen, L.N. Cooper and J.R. Schrieffer, *Phys. Rev.*, 1957, **108**, 1175.

46 W.A. Little, *Phys. Rev.*, 1964, **A134**, 1416.

47 D. Jerome, A. Mazaud, M. Ribault and K. Bechgaard, *J. Phys. Lett.*, 1980, **41**, L95.

48 K.D. Karlson, G.W. Crabtree, L. Nunez, H.H. Wang, M.A. Beno, U. Geiser, M.A. Firestone, K.S. Webb and J.M. Williams, *Solid State Comm.*, 1986, **57**, 89.

49 J.D. Wright, *Prog. Surface Sci.*, 1989, **31**, 1.

50 F. Gutmann and L.E. Lyons, *Organic Semiconductors*, New York: Wiley, 1967.

51 R.L. van Ewyk, A.V. Chadwick and J.D. Wright, *J. Chem. Soc. Faraday I*, 1980, **76**, 2194.

52 B. Rosenberg, T.N. Misra and R. Switzer, *J. Chem. Phys.*, 1968, **48**, 2096.

53 R.J. Cherry and D. Chapman, *Nature*, 1967, **215**, 956.

54 B. Bott and T.A. Jones, *Sensors and Actuators*, 1984, **5**, 43.

55 C.J. Schramm, R.P. Scaringe, D.R. Slojakovic, B.M. Hoffman, J.A. Ibers and T.J. Marks, *J. Am. Chem. Soc.*, 1980, **102**, 6702.

56 A.V. Chadwick, P.B.M. Dunning and J.D. Wright, *Mol. Cryst. Liq. Cryst.*, 1986, **134**, 137.

57 P.B.M. Archer, A.V. Chadwick, J.J. Miasik, M. Tamizi and J.D. Wright, *Sensors and Actuators*, 1989, **16**, 379.

58 A.V. Chadwick, A. Wilson and J.D. Wright, *Sensors and Actuators*, 1991, **B4**, 499.

59 A. Wilson, G.P. Rigby, J.D. Wright, S.C. Thorpe, T. Terui and Y. Maruyama, *J. Mater. Chem.*, 1992, **2**, 303.

60 P. Roisin, J.D. Wright, R.J.M. Nolte, O.E. Sielcken and S.C. Thorpe, *J. Mater. Chem.*, 1992, **2**, 131.

61 J.D. Wright, P. Rosin, G.P. Rigby, R.J.M. Nolte, M.J. Cook and S.C. Thorpe, *Sensors and Actuators*, 1993, **B13**, 276.

62 J.W. Weigl, *Angew. Chemie (Int. Edn)*, 1977, **16**, 374.

63 G.A. Chamberlain, *Solar Cells*, 1983, **8**, 47.

64 M.S. Wrighton, *Acc. Chem. Res.*, 1979, **12**, 303.

65 V.M. Vincent and J.D. Wright, *J. Chem. Soc. Faraday I*, 1974, **70**, 58.

66 R.R. Chance and C.L. Braun, *J. Chem. Phys.*, 1973, **59**, 2269; 1976, **64**, 3573.

67 L. Onsager, *Phys. Rev.*, 1938, **54**, 554.

68 J. Noolandi and K.M. Hong, *J. Chem. Phys.*, 1979, **70**, 3230.

69 Z.D. Popovic, *J. Chem. Phys.*, 1983, **78**, 1552.

70 B.J. Mulder, *Philips Res. Rep. Suppl.*, 1968, **4**, 44.

71 R.L. van Ewyk, A.V. Chadwick and J.D. Wright, *J. Chem. Soc. Faraday I*, 1981, **77**, 73.

72 D.D. Eley, E. Metcalfe and P. White, *J. Chem. Soc. Faraday I*, 1975, **71**, 955.

73 S.M. Ryvkin, *Soviet Phys. – J. Exper. Theor. Phys.*, 1950, **20**, 139.

74 P. Gregory, *High-technology Applications of Organic Colorants*, New York: Plenum, 1991.

75 H. Böttcher, T. Fritz and J.D. Wright, *J. Mater. Chem.*, 1993, **3**, 1187.

9

Special topic: C_{60}

In 1987 when this book was first published, the third form of carbon, C_{60}, had been discovered for barely two years and was known only as minute traces formed by laser evaporation of graphite and detectable using mass spectrometry. Such is the pace of modern science that in the subsequent six years C_{60} has not only been isolated in substantial quantities but is also available commercially from several suppliers, and its chemistry, structural, optical and electrical properties have been extensively explored. The discoveries that C_{60} forms superconducting compounds with transition temperatures significantly higher than the charge-transfer salts discussed in chapter 8 as well as p/n junction devices with potential application in solar cells and photodetectors, together with the sheer fascination of this novel material, have led to this branch of the study of molecular crystals receiving wide media coverage. In fact, C_{60} provides examples of many of the topics covered in other parts of this book and is, therefore, included as one of the special topics in this edition.

9.1   Formation and structure
The structure of C_{60} was first suggested in 1970 by Osawa in Japanese language publications which received little attention outside Japan at the time[1]. In 1984[2] mass spectra of carbon clusters produced by laser vaporisation of graphite in a helium atmosphere showed that many clusters of more than 40 carbon atoms were detectable. Only even-numbered clusters were stable, and the abundance of C_{60} and to a lesser extent C_{70} suggested that these species were particularly stable. In 1985 it was discovered that by optimising the conditions of evaporation, C_{60} could be made by far the dominant species produced, and the model shown in figure 9.1 was proposed for the structure of the molecule[3]. In 1990 an

optimised method of preparing macroscopic samples of C_{60} by vaporising graphite in a carbon-arc discharge in an atmosphere of approximately 100 mbar helium was reported, and the discovery was made that benzene extraction of the dust produced by this process gave a purple solution from which small crystals could be isolated. These consisted of a mixture of C_{60}, C_{70} and small traces of other species[4]. Chromatography of a hexane solution on alumina separates C_{60} and C_{70}, the separation improving at higher temperatures. In 1991 it was shown that these molecules can also be isolated from the flames of benzene burning in air, with yields up to 2.8 g of $(C_{60} + C_{70})$ per kg of benzene burned. Interestingly, the largest yields do not occur in the most heavily sooting flames, and the yield increases as temperature increases or pressure decreases (whereas total soot yields decrease in these conditions)[5]. Such observations led Kroto to comment that the material 'has been under our noses since time immemorial and yet it was only discovered in 1985 and isolated and its structure confirmed as recently as 1990'[6].

Much attention has been given to the mechanism by which vaporised carbon forms such closed structures[7]. Vaporisation is believed to lead to the initial formation of linear chains of carbon atoms with reactive ends which link to form cyclic structures and eventually polycyclic networks. These networks have many dangling bonds and normally are so reactive that they link rapidly, forming large soot particles. However, if they can be maintained at high temperature without such reactions they have a marked tendency to curl so as to form near-spherical structures in which dangling bonds on the *same* sheet can be eliminated by linking with each

9.1.

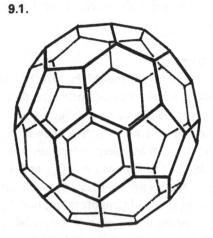

other, thus minimising energy. Euler showed[8] that such spheroidal structures can be constructed from varying numbers of hexagons provided there are always 12 pentagons present. Thus, C_{60} can be regarded as formed from 20 hexagons + 12 pentagons (see figure 9.2 for a template to construct a three-dimensional model), while C_{70} consists of 25 hexagons + 12 pentagons. The structure of C_{60} (figure 9.1) has the high rotational symmetry of an icosahedron, with six 5-fold axes, ten 3-fold axes, fifteen 2-fold axes and inversion symmetry. Such structures have been used in several notable geodesic domes designed by the American architect Buckminster Fuller, and this has produced the unique situation in which one form of a chemical element has been named after an architect: buckminsterfullerene. (This rather long name is nevertheless infinitely preferable to the IUPAC alternative[9]: hentriacontacyclo[$29.29.0.0^{2,24}.0^{3,12}.0^{4,59}.0^{5,10}.0^{6,58}.0^{7,55}.0^{8,53}.0^{9,21}.0^{11}$ $^{,20}.0^{13,18}.0^{15,30}.0^{16,28}.0^{17,25}.0^{22,52}.0^{23,50}.0^{26,49}.0^{27,47}.0^{29,45}.0^{32,44}.0^{33,60}$ $.0^{34,57}.0^{35,43}.0^{36.56}.0^{37,41}.0^{38,54}.0^{39,51}.0^{40,48}.0^{42,46}$]hexaconta1,3,5(10), 6,8,11(18),14,16,19,21,23,25,27,29(45),30,32(44),33,35(43),36,38(54),39 (51),40(48),41,46,49,52,55,57,59-triacontaene.) Other compounds of this general class are now commonly referred to as fullerenes.

9.2.

9.2   Crystal structure and phase transitions

As predicted from the packing considerations discussed in section 3.1, C_{60} forms an orientationally disordered fcc structure at room temperature[10]. Solid-state ^{13}C NMR spectroscopy at room temperature shows a single sharp line at 143 ppm as expected for rapid isotropic re-orientation. On cooling, a phase transition to an orientationally ordered simple cubic structure occurs at 249 K, as shown by X-ray diffraction and calorimetric studies[11]. At 283 K, the NMR rotational correlation time τ is 9.2 ps, which is only three times longer than estimated for the gas-phase molecule and significantly shorter than that of 15.5 ps for the solution of C_{60} in 1,1,2,2-tetrachloroethane at the same temperature. Studies of the temperature-dependence of this correlation time for solid C_{60} show[12] a sharp increase in τ below the phase transition, accompanied by a change in the slope of the plot of $\ln\tau$ versus $1/T$ corresponding to a change in the activation energy for reorientation from 5.9 to 19 kJ mol^{-1}. The low-temperature phase is therefore believed to correspond to a state in which the molecules perform activated jumps between symmetry-equivalent orientations, often referred to as the 'ratchet' phase. These re-orientations make it very difficult to use X-ray diffraction measurements to obtain detailed information on the molecular geometry (although, very recently, refinement of low-temperature diffraction data on C_{60} and its solvates $C_{60}.C_6H_6.CH_2I_2$ and $C_{60}.4C_6H_6$ in terms of disordered models has yielded interatomic distances[13]). Scanning tunnelling microscopy has also been used to confirm fcc packing of films of C_{60} grown on GaAs(110) substrates, but, as with X-ray diffraction, the molecular rotation precludes the possibility of ever being able to 'see' individual carbon atoms in C_{60} using this technique[14]. However, it has proved possible to determine the bond lengths in C_{60} using solid-state NMR on a sample enriched to 6% with ^{13}C, using the fact that the magnetic dipolar coupling of ^{13}C–^{13}C depends on the inverse cube of the C–C distance and leads to a quite sharp 'Pake' doublet[15] in the spectrum. There are two types of C–C bond in the structure shown in figure 9.1: 60 pentagon edges and 30 links between pentagons. Spectral simulation gave a good fit to the observed data when these two types of bond had lengths of 1.45 ($\pm$0.015) and 1.40 ($\pm$0.015) Å, respectively. From these values, the diameter of C_{60} can be calculated to be 7.1 $\pm$ 0.07 Å[16]. These results provide an excellent example of the value of NMR in the study of molecular crystals.

Under external pressure, these structural features lead to some interest-ing properties[17]. The compressibility along any axis is comparable to that of graphite along its inter-sheet c-axis since, like graphite, the structural units are held together only by weak van der Waals forces. The molecules themselves retain the same size during compression, which only alters the intermolecular distance, and it has been estimated that if they could be squeezed together to occupy only 70% of their initial volume the resulting material would be even harder than diamond. Indeed, experiments in which C_{60} molecules are impacted onto a steel surface at $28\,000\,km\,h^{-1}$ leave the spherical structures intact, showing their immense resilience. Reducing the intermolecular separation by compression significantly increases the activation energy for reorientation in the solid, so that even at a pressure of 3 kbar the transition between rotationally disordered and ordered phases already occurs above room temperature.

9.3 Electronic structure

The electronic structure[18] of C_{60} is unique for a π-bonded hydrocarbon, in that the molecule is a strong electron acceptor with an electron affinity of $2.65\,eV$[19]. This is a direct consequence of its geometric structure, which influences the electronic energy levels in two quite simple ways. First, as mentioned above, the structure may be regarded as 20 six-membered rings with 12 five-membered rings. Conjugated five-membered rings always lead to higher electron affinity as a result of the aromatic stability associated with the cyclopentadienyl anion (c.f. the somewhat similar arguments in section 8.6 favouring the enhanced donor power of TTF because of the gain of aromaticity on loss of an electron from the five-membered rings containing two sulphur atoms). Second, the non-planarity of the structure means that the π-system electrons are no longer pure p in character. The slight pyramidalisation of each carbon atom induces a small re-hybridisation in which some s-character is introduced into the π-orbitals. Figure 9.3 shows the estimated degree of s-character as a function of the pyramidalisation angle (defined as the mean downward angular deflection of the three atoms surrounding each carbon from the plane in which they would all lie in the graphite structure). For C_{60} this angle is $11.6°$ which, from figure 9.3, would correspond to nearly 10% of 2s character. Although this value is much higher than an experimental estimate of 3% obtained from an analysis of ^{13}C NMR coupling constants, it serves to illustrate the importance of the effect. Figure 9.4 shows the resulting molecular orbital energy level diagram for C_{60}, from which it can be seen that the molecule might be expected to accept at least six electrons, and possibly up to 12, in the right conditions. In fact as many as six separate reversible electrochemical redox waves have recently been observed in cyclic voltammetry of C_{60} at $-10°C$ in toluene/CH_3CN solution[20].

9.4   Alkali metal salts

The high electron affinity of C_{60} explains observations reported in 1987 on the concentration profiles of negative polyhedral ions in benzene flames[21], which showed that C_{60} monoanions formed at shorter distances from the burner than other carbon cluster anions and persisted far longer as the ions move further from the burner. A natural development has also been to explore the formation of charge-transfer complexes with organic donors, including conducting polymers, and of salts (both with organic cations and with simple metal cations). Thin films of C_{60} changed colour from yellow to magenta when exposed to alkali metal vapours. During the exposure the electrical conductivity of the films at room temperature first increased by several orders of magnitude and then decreased. From the molecular orbital diagram in figure 9.4 it might be expected that metallic character would be observed as electrons are added to the lowest unoccupied molecular orbital (t_{1u}), reaching a maximum when this is half-filled (formally the 3– anion) and eventually falling back to something near the original poor semiconductor value when sufficient electrons have been added to form the 6– ion[18]. The actual process of doping is also influenced by the positions available in the three-dimensional solid structure for the metal cations. Synchrotron radiation X-ray diffraction

9.3. The s-character of the π-orbital as a function of the pyramidalisation angle.

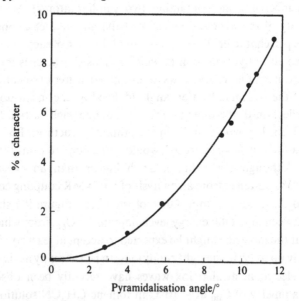

studies on the thin films have been valuable in determining the structures which are adopted in K_3C_{60}, Rb_3C_{60}, K_4C_{60}, K_6C_{60}, Rb_6C_{60} and Cs_6C_{60}. The structures of the compounds containing three alkali metal cations per C_{60} are fcc like the parent hydrocarbon, but with alkali metal ions in each of the (one octahedral and two tetrahedral) sites per C_{60}^{3-} species. Clearly such a structure can only accommodate up to three cations per C_{60}, so it is not surprising that the compounds with six alkali cations per C_{60} have a bcc structure, with the face centres occupied by four cations approximately in a square. Although some interesting structures might be expected for doping with less than three cations per anion, where choices would be available for occupation of octahedral and tetrahedral sites, so far it appears that there is

9.4. Molecular orbital energy level diagram for C_{60}.

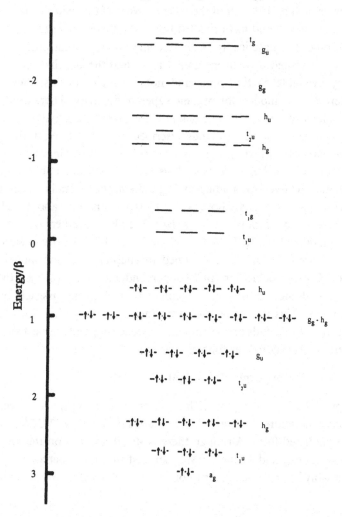

a marked tendency for such low-doped materials to disproportionate to yield separate phases of the M_3C_{60} and pure C_{60} species. Using the electron-reservoir complex $Fe^I(C_5H_5)(C_6(CH_3)_6)$, three very air-sensitive compounds containing C_{60} $1-$, $2-$ and $3-$ ions have been isolated, with the $2-$ ion showing an intrinsic magnetic state[22]. K_4C_{60} forms a body-centred orthorhombic structure and although it would be expected to be metallic with a part-filled band in fact it has been found to be a poor conductor. It has been suggested that this may be a consequence of a Jahn–Teller distortion which would open up a gap in the band structure[18]. The most dramatically interesting feature of these materials is the superconductivity which has been observed in the compounds of C_{60}^{3-}, Rb_2CsC_{60} having a superconducting transition temperature of 33 K while the Rb/Tl doped material has been claimed to have a transition temperature as high as 45 K. Until the recent advent of the oxide semiconductors, these values would have rivalled the best known for inorganic materials. Although the practicality of these compounds as electronic materials is severely limited by their instability in air and the fact that they are so far only available as thin films and not as single crystals, they are very interesting as models for organic superconductivity. There are two clear reasons why they are better candidates for superconductivity as a class than the charge-transfer compounds described in chapter 8. First, the spherical symmetry of C_{60} favours cubic or near-cubic structures in which anisotropy and one-dimensional effects such as Peierls distortions are not favoured. Second, the overlap of adjacent C_{60} units in the crystalline solid is small, giving narrow bands whose width (~ 0.5 eV) is only about half that in complexes of donors such as BEDT-TTF. Together with the triple degeneracy of the t_{1u} LUMO, this leads to a high density of states at the Fermi level, $N(E_F)$. As mentioned in chapter 8, in the model of Bardeen–Cooper–Schrieffer (BCS) superconductivity arises from the highly organised motion of pairs of conduction electrons assisted by lattice phonons. This theory predicts that the superconducting transition temperature T_c depends on the phonon frequency ω_{ph} and the electron-phonon coupling strength V as follows:

$$T_c = \omega_{ph} \exp(-1/VN(E_F)) \tag{9.1}$$

Figure 9.5 shows a plot of $\ln(T_c)$ versus $1000/N(E_F)$ for several such superconductors, (where $N(E_F)$ is in units of states $eV^{-1} C_{60}^{-1}$), which has the predicted form. Although there is still discussion on the appropriate values of ω_{ph} and V, these results suggest that the superconductivity can probably be explained in terms of the BCS theory with less difficulties than in the case of other classes of organic superconductor. Furthermore, the idea that the transition temperature is related to the density of states at the Fermi level and hence directly related to the lattice spacing in a series of M_3C_{60} compounds has also been shown to be correct and independent of the nature of M. Figure 9.6 shows T_c and lattice parameter (a) values both for a series of measurements at different pressures on the potassium and rubidium compounds and for a series of measurements at ambient pressure on compounds with different combinations of alkali metals, and the points lie reasonably well on a single curve as predicted.

9.5   Salts with organic cations

From the results on potassium and rubidium compounds, it might be expected that particularly interesting compounds would result from the combination of a large organic strong electron-donor molecule and C_{60}. Tetrakis dimethylaminoethylene (TDAE) is such a donor, and forms a 1 : 1 compound with C_{60}. However, this compound crystallises in a C-centred monoclinic lattice in which C_{60} molecules are separated by 9.28 Å along the c-axis but by 10.25 Å in the ab plane[23]. Despite the very large size of the TDAE cation, these distances are comparable to those (10.07–10.25 Å) in alkali-metal-doped samples, so this approach does not lead to a narrower

9.5.

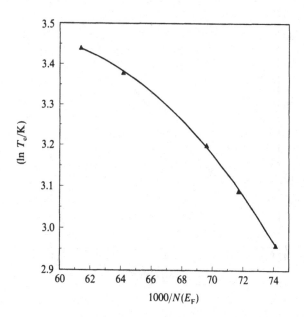

band and the hoped-for higher density of states at the Fermi level. Furthermore, the crystal structure leads to an anisotropic band structure, because the overlap along the *c*-axis is different from that in the *ab* plane. This anisotropy favours Peierls distortions, so the material is not in fact a useful conductor. However, on cooling below 17.5 K the compound undergoes a ferromagnetic transition and measurements with a Faraday balance indicate one radical spin per molecule. A new, broad, intense ESR signal below 17.5 K suggests that radical spins in the ferromagnetically ordered state are localised on the C_{60} units[24].

9.6   Applications in junction devices

C_{60} has also been used in a semiconductor junction device with potential application as a solar photovoltaic cell or as a photodetector[25]. A sandwich cell consisting of a layer of C_{60} on a gold electrode, covered with a layer of the polymer poly[2-methoxy,5-(2'-ethylhexyloxy)-*p*-phenylenevinylene] (MEHPPV) with a transparent indium tin oxide top electrode gives a photovoltaic response for photon energies from 1.7–4 eV, with a narrow window of reduced response around 2.5 eV where the polymer itself absorbs strongly. As a photocell, such a device gives a linear current

9.6. Dependence of T_c on lattice parameter (*a*) for M_3C_{60} compounds.

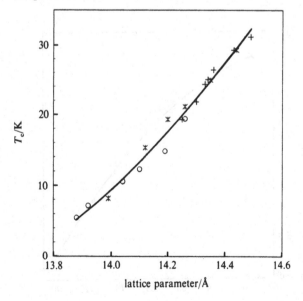

response for incident light intensities ranging from 10^{-5}–1 W cm^{-2}. Since both C_{60} and the polymer are soluble, these devices can be made by spin coating at low cost, and their mechanical flexibility and light weight offer significant advantages.

9.7   Chemistry of C_{60}

The electronic and geometrical structure of C_{60} combine to provide three distinctive chemical features: (i) the high electron affinity discussed above, together with the π-bonded structure, leads to numerous reaction possibilities with nucleophiles; (ii) the distortion of the bonds around each carbon atom from planarity means that C–C bonds are effectively already in an ideal geometry favouring η^2 bonding to transition metals in low valency states; (iii) the unique spherical structure imposes packing limitations on the number of sites at which reaction can take place. A simple view of the C_{60} molecule is that it is a spherical assembly of six linked fulvalene units (figure 9.7) with a strong tendency for the two five-membered rings to pick up electrons and thus gain aromaticity, with the linking double bond becoming single bond in character. If each of these double bonds coordinated to a transition metal, an effectively octahedral structure would be formed around the C_{60} core, with the transition metal groups located as far apart as possible from each other. These features are clearly reflected in the chemistry of C_{60}, which is now beginning to emerge as its production on the scale required for synthetic chemistry is increasingly accessible.

One example of a typical reaction is the addition of diphenyldiazomethane yielding C_{61}(phenyl)$_2$. This type of reaction initially leads to at least two isomeric products, as shown by the appearance of distinct ^{13}C NMR resonances in regions appropriate for the extra bridging carbon, but

9.7.

after refluxing in a suitable solvent overnight only a single isomer is observed, corresponding to addition across the double bond joining two five-membered rings. This is the thermodynamically most stable isomer. The addition of an extra carbon atom to the shell, with associated substituents, leads to non-spherical symmetry and hence limitations on molecular rotation in the solid state. Even the simplest species of this type, $C_{61}H_2$, shows preferential rotation about one axis which passes through the extra carbon atom. A natural extension to this type of chemistry is to functionalise the substituents on the extra carbon so as to permit the linking of two or more substituted C_{61} units into dimers, oligomers or even polymers. Knowledge of the chemistry of these species is expanding rapidly.

Transition metal complexes of C_{60} can be prepared simply by direct substitution reactions. Thus, addition of $[(C_6H_5)_3P]_2Pt(\eta^2\text{-}C_2H_4)$ to a magenta-coloured toluene solution of C_{60} at 25 °C leads to an instantaneous colour change to dark green, with a 75% yield of the complex $[(C_6H_5)_3P]_2Pt(\eta^2\text{-}C_{60})$. Similarly, direct reaction of $(Et_3P)_4M$ (where $M = Ni, Pd, Pt$) with C_{60} yields either green 1 : 1 compounds or orange-red $[(Et_3P)_2M]_6C_{60}$. Examination of space-filling models shows that in the latter complexes the phosphine ligands effectively occupy all of the available space around the carbon core, so no higher coordination-number complexes are possible.

These materials have not yet found practical applications, although clearly they have potential as electronic and non-linear optical materials. In the long term it may well be that the greatest significance of the discovery of C_{60} lies not just in the attractiveness of its own properties but rather in the wide range of novel materials for which it is a starting material. In this context, it is perhaps appropriate to end with a warning that all such radically new compounds should be handled carefully until their health hazards have been established. Future potential with harmful consequences must never be overlooked in the quest for specific applications potential, and the known hazards of polycyclic aromatic hydrocarbons suggest that such possibilities certainly require investigation for these new compounds.

References

1 E. Osawa, *Kagaku (Kyoto)*, 1970, **25**, 854. Z. Yoshida and E. Osawa, *Aromaticity*, Kagaku Dojin: Kyoto, 1971.

2 E.A. Rohlfing, D.M. Cox and A. Kaldor, *J. Chem. Phys.*, 1984, **81**, 3322.

3 H.W. Kroto, J.R. Heath, S.C. O'Brien, R.F. Curl and R.E. Smalley, *Nature*, 1985, **318**, 162.

4 W. Krätschmer, L.D. Lamb, K. Fostiropoulos and D.R. Huffman, *Nature*, 1990, **347**, 354.

5 J.B. Howard, J.T. McKinna, Y. Makarovsky, A.L. Lafleur and M.E. Johnson, *Nature*, 1991, **352**, 139.

6 H.W. Kroto, *J. Chem. Soc. Faraday I*, 1991, **87**, 2871.

7 R.E. Smalley, *Acc. Chem. Res.*, 1992, **25**, 98.

8 L. Euler, 1707–83. This result is related to the better-known *Euler Theorem*, that for a closed polyhedron the sum of the number of vertices and faces is equal to the number of edges plus 2.

9 P. Rose, personal communication to H.W. Kroto, quoted as reference 47 in H.W. Kroto, A.W. Allaf and S.P. Balm, *Chem. Rev.*, 1991, **91**, 1213.

10 H.-B. Bürgi, E. Blanc, D. Schwarzenbach, S. Liu, Y.-J. Lu, M.M. Kappes and J.A. Ibers, *Angew. Chem. (Int. Edn)*, 1992, **31**, 640.

11 P.A. Heiney, J.E. Fischer, A.R. McGhie, W.J. Romanow, A.M. Denenstein, J.P. McCauley Jr and A.B. Smith III, *Phys. Rev. Lett.*, 1991, **66**, 2911.

12 R. Tycko, G. Dabbagh, R.M. Fleming, R.C. Haddon, A.V. Makhija and S.M. Zahurak, *Phys. Rev. Lett.*, 1991, **67**, 1886.

13 A.L. Balch, J.W. Lee, B.C. Noll and M.M. Olmstead, *J. Chem. Soc. Chem. Comm.*, 1993, 56.

14 J.H. Weaver, *Acc. Chem. Res.*, 1992, **25**, 143.

15 G.E. Pake, *J. Chem. Phys.*, 1948, **16**, 327.

16 C.S. Yannoni, P.P. Bernier, D.S. Bethune, G. Meijer and J.R. Salem, *J. Am. Chem. Soc.*, 1991, **113**, 3190.

17 G. Kriza, J.-C. Ameline, D. Jerome, A. Dworkin, S. Szwarc, C. Fabre, D. Schutz, A. Rassat and P. Bernier, *J. Phys. I France*, 1991, **1**, 1361. G.A. Samara, J.E. Schriber, B. Morosin, L.V. Hansen, D. Loy and A.P. Sylwester, *Phys. Rev. Lett.*, 1991, **67**, 3136.

18 R.C. Haddon, *Acc. Chem. Res.*, 1992, **25**, 127.

19 L.S. Wang, J. Conceicao, C. Jin and R.E. Smalley, *Chem. Phys. Lett.*, 1991, **182**, 5.

20 Q. Xie, E. Pérez-Cordero and L. Echegoyen, *J. Am. Chem. Soc.*, 1992, **114**, 3978.

21 Ph. Gerhardt, S. Löffler and K.H. Homann, *Chem. Phys. Lett.*, 1987, **137**, 306.

22 C. Bossard, S. Rigaud, D. Astruc, M.-H. Delville, G. Félix, A. Février-Bouvier, J. Amiell, S. Flandrois and P. Delhaes, *J. Chem. Soc. Chem. Comm.*, 1993, 333.

23 P.W. Stephens, D. Cox, J.W. Lauta, L. Milhaly, J.B. Wiley, P.M. Allemand, A. Hirsch, K. Holczer, O. Li, J.O. Thompson and F. Wudl, *Nature*, 1992, **355**, 331.

24 K. Tanaka, A.A. Zakhidov, K. Yoshizawa, K. Okahara, T. Yanabe, K. Yakushi, K. Kikuchi, S. Suzuki, I. Ikemoto and Y. Achiba, *Phys. Lett. A*, 1992, **164**, 221.

25 N.S. Sacriciftcy, D. Braun, C. Zhang, V.I. Srdanov, A.J. Heeger, G. Stucky and F. Wudl, *Appl. Phys. Lett.*, 1993, **62**, 585.

10

Special topic: non-linear optic materials

When light passes through a dielectric medium, the associated oscillating electric field of the light induces an oscillating polarisation, P, in the medium. This in turn re-radiates an oscillating electric field, the sum of the re-radiated oscillating fields from all illuminated regions constituting the propagated light beam in the medium. For low light intensity and a totally isotropic medium the induced polarisation is related to the electric field E by

$$P = \varepsilon_0 \chi E \tag{10.1}$$

where ε_0 is the permittivity of free space and χ is the polarisability of the medium. χ is related to the refractive index (n):

$$n^2 = 1 + \chi. \tag{10.2}$$

For higher light intensities, such a simple linear response is no longer observed and, in a manner analogous to the onset of anharmonicity in vibrational spectroscopy as the amplitude of vibration increases, terms in higher powers of the electric field must be considered:

$$P = \varepsilon_0 \{ \chi^{(1)} E + \chi^{(2)} E^2 + \chi^{(3)} E^3 + \ldots \}. \tag{10.3}$$

The terms $\chi^{(2)}$, $\chi^{(3)}$, etc. are known as the second-, third-, (etc.) order hyperpolarisabilities. It is these terms in higher powers of electric field which lead to the name 'non-linear optics' (NLO).

Materials which show large non-linear optical properties are of considerable current interest for optical and electro-optical applications. Several reasons for this are soon obvious from a consideration of the implications of equations 10.1–10.3. Thus, rewriting equation 10.2 as

$$n^2 = 1 + \chi_{\text{eff}} \tag{10.4}$$

where $\chi_{\text{eff}} \sim \chi^{(1)} + \chi^{(2)}E$, gives:

$$n^2 \approx 1 + \chi^{(1)} + \chi^{(2)}E \tag{10.5}$$

so that the refractive index of the medium is now dependent on the electric field. Therefore, application of an external electric field can be used to modulate the amplitude or phase of the light beam, converting an electrically carried signal into one carried optically (and thus faster and with no electrical interference). Furthermore, if third-order terms are considered, equation 10.5 must include a term in E^2 and the refractive index becomes dependent on the light intensity, providing switching possibilities. For example, optical waveguides or optical fibres depend on guiding light by total internal reflection at boundaries between two media of different refractive index. If the refractive index of one of these media can be changed, either by application of an external electric field or by increasing the intensity of the light itself, until it is close to that of the other material, total internal reflection will no longer occur and the light will be switched out of the guided mode. In a much more subtle application, two intense input ('pump') beams of the same frequency are allowed to interfere within a third-order NLO material, producing a periodic variation in refractive index corresponding to intense and dark regions of the interference pattern. This acts as a diffraction grating for a third ('probe') laser beam, and the deflection angle of this beam is therefore related to the frequency of the interfering beams. Changes in the frequency of the pump beams, therefore, are very rapidly converted to changes in position of the diffracted probe beam, which may be determined using, for example, a diode array, with applications in signal processing. This effect is known as *degenerate four-wave mixing*.

Consider also the fact that a coherent laser beam of frequency v has an associated time-dependent electric field of amplitude $E_v \sin(2\pi vt)$, so that the polarisability including up to second-order terms is given by:

$$
\begin{aligned}
P &\approx \varepsilon_0(\chi^{(1)}E + \chi^{(2)}E^2) \\
&\approx \varepsilon_0(\chi^{(1)}E + \chi^{(2)}E_v{}^2\sin^2(2\pi vt)) \\
&\approx \varepsilon_0(\chi^{(1)}E + \tfrac{1}{2}\chi^{(2)}E_v{}^2(1 - \cos 4\pi vt)).
\end{aligned}
\tag{10.6}
$$

Therefore, the induced oscillating polarisation has a component of double the input frequency, and part of the propagated light is frequency-doubled. This second harmonic generation (SHG) is of particular importance in increasing the quantum energy available from a laser source by producing light of half the wavelength. For example, the information density on a compact disc is ultimately limited by diffraction effects in the optical system, and halving the laser wavelength would increase the information storage density by a factor of four. In a similar way, it can be shown that third harmonic generation may also occur, although with lower efficiency because of the smaller third-order coefficients.

A related phenomenon is frequency mixing, in which *two* input beams of *different* frequencies interact in the material to produce outputs of frequencies equal to the sum and difference of those of the two inputs. This is a useful way of shifting the frequency of a signal-carrying modulated light beam into a region where detection is easier. For example, the availability of lasers and good optical fibres for infra-red frequencies means that information transfer is often done in this region of the spectrum, whereas detection is easier in the visible region where higher photon energies are available. The reverse process, where an input beam is split into two lower frequencies whose values are determined by the refractive indices and orientation of the NLO material with respect to the direction of the incident beam, is also possible and is useful for providing tunable laser sources.

These and other applications of NLO properties have been widely explored using inorganic materials such as lithium niobate and potassium dihydrogen phosphate (KDP), although the former is relatively difficult to produce and is damaged by high laser intensities while the latter has quite small non-linear coefficients. Molecular crystals offer several distinct advantages over inorganic materials for NLO applications:

(1) *Their non-linear coefficients can be substantially higher than those for inorganic materials*, because of their high electronic polarisability particularly where π-electron systems (especially those linking electron-donor groups with electron-acceptor groups) are present. For example, 4-(N,N-dimethylamino)-3-acetamidonitrobenzene (DAN) (figure 10.1) is about 100 times more effective for second

10.1. DAN.

harmonic generation than KDP[1]. This factor is particularly significant for applications involving frequency-doubling of relatively low-powered laser diodes of the type used in many current optical devices.

(2) *The non-linear responses can be very rapid because the induced polarisation arises from movement of electrons rather than nuclei.*

(3) *The stability of organic materials under illumination by intense laser beams can be higher than that of inorganics.* This has been established by experiment (e.g. the optical damage threshold for urea under 10 ns pulses of 1.06 μm laser radiation is 5 GW cm^{-2} compared with 0.5 GW cm^{-2} for KDP[2]), and is generally believed to be a consequence of the availability of many modes of molecular and lattice motions/vibrations into which the energy of any absorbed light may be dispersed, thus limiting local damage. (However, long-term stability in devices has in general not been established yet for this relatively new class of materials. It would be surprising to find very good long-term stability in such applications as SHG where high-energy quanta are available and likely to cause photochemical damage by a variety of mechanisms over a period of time.)

(4) *The NLO properties of single molecules (i.e. molecular hyperpolarisabilities, see below) can be calculated using molecular orbital methods.* The results of such calculations are in good agreement with experimental results from solutions of the molecules in question and permit preliminary screening of candidate molecules and elimination of unpromising materials before chemical synthesis and crystal growth programmes are started[3]. However, this approach cannot predict the NLO properties of the solid material, which depend on local electric fields within the crystal as well as on the crystal structure. For example, very accurate X-ray diffraction work has permitted determination of atomic charges and hence actual dipole moments of polar molecules within crystals and has shown that these are substantially higher than those of the isolated molecules because of the co-operative effects of surrounding molecules on the local electric field[4]. Furthermore, both the molecular and crystal hyperpolarisabilities are in fact not isotropic as assumed in the above equations, and both the anisotropy introduced by the molecular structure and that introduced by the crystal structure must be considered when describing NLO effects of molecular crystals.

(5) *The anisotropy of organic molecules means that molecular crystals frequently show high birefringence, which can be used to obtain 'phase matching' to maximise the intensity of frequency-doubled light.* The problem of phase matching arises because most materials have refractive indices dependent on the wavelength of light. As incident light traverses a crystal, frequency-doubled components are generated all along the light path. Since the velocity of these components is not the same as that of the incident wave, they cannot all be in-phase and interference effects occur, reducing the intensity of the 'second harmonic' light. Now, because both molecular and crystal hyperpolarisabilities are in fact anisotropic, it is possible for the extraordinary-polarised incident wave to generate a second harmonic with ordinary polarisation. Then by adjusting the angle of incidence of the laser beam onto the crystal until the extraordinary refractive index is exactly equal to the ordinary refractive index for the frequency-doubled component, both will travel with the same velocity and no interference effects will be observed. This condition is known as phase-matching[5]. Details of how such an optimum angle of incidence is calculated are beyond the scope of this text. The important feature from a materials viewpoint is molecular and crystal anisotropy, which is much easier to achieve, in molecular crystals than in inorganic solids.

10.1   Molecular design for high second-order NLO coefficients

On a molecular scale, an equation corresponding to equation 10.3 for bulk material can be written, relating the induced dipole moment, μ, to the molecular polarisability, α, and hyperpolarisabilities, β, γ, etc. as follows:

$$\mu = \alpha E + \beta E^2 + \gamma E^3 + \ldots \tag{10.7}$$

A high molecular hyperpolarisability implies that the molecule changes its electron distribution easily, i.e. has a low-lying excited state with a dipole moment substantially different from the ground state. In chemical terms this requires a facile intramolecular charge-transfer process, and this is clearly achieved by substituting an aromatic or other conjugated molecule with an electron-donor group on one end and an electron-acceptor group on the other end. Thus, for example, the molecule *p*-nitroaniline might at first sight appear a suitable candidate. However, for a solid material the second-order and other even-order hyperpolarisabilities are zero if there is a centre of symmetry, and, as discussed in chapter 3, dipolar molecules such as l-nitroaniline commonly tend to pack with centrosymmetric anti-parallel molecular alignment (see, for example, the structure of *p*-iodoben-zonitrile). Thus in practice, *p*-nitroaniline is of no use as a second-order non-linear optic material in the solid state. It is therefore necessary to design molecules which not only possess the necessary donor and acceptor groups but also form non-centrosymmetric solids. This can be achieved in three ways.

Addition of substituents to prevent centrosymmetric packing
The simple example of this approach involves adding a methyl group to *p*-nitroaniline to give 2-methyl-4-nitroaniline (MNA) (figure 10.2). Al-though the methyl group is not very large it is sufficient to occupy space normally required by the antiparallel molecule, forcing a non-centrosym-metric structure. In the similarly shaped molecule 3-methyl-4-nitropyridine *N*-oxide (POM) (figure 10.3) the ground state dipole moment is near-zero, which further discourages a centrosymmetric lattice packing. Alternatively, deliberate introduction of additional strong dipolar interactions such as hydrogen bonding to dominate the natural anti-parallel alignment of the rest of the molecule can be an effective strategy. Random substitution of this type cannot of course be expected to succeed in every case and it has been estimated that an entirely empirical approach leads statistically to only 29% of acentric crystals, of which only a few would have the desired NLO efficiency[6]. Furthermore, the problem of polymorphism makes it very difficult to predict the crystal structure actually obtained with complete certainty. A related approach guarantees a non-centrosymmetric solid by using a chiral group as the substituent. For example, replacing the NH_2

10.2. MNA.

10.3. POM.

group in *p*-nitroaniline by $(PhCH_2)(HOCH_2)CH-NH-$ generates a $P2_1$ spacegroup in which the second-harmonic generation efficiency is 150 times that of urea while the molecules happen to be aligned almost ideally for phase-matching as described above[6]. However, although chirally substituted molecules must crystallise in non-centrosymmetric structures, this is not a *necessary* guarantee that the materials will have high $\chi^{(2)}$ values and crystals of some chiral compounds are known to show near-zero second-harmonic generation efficiency.

Use of Langmuir–Blodgett films

As discussed in chapter 3, the Langmuir–Blodgett technique can be used to prepare films in which ordered layers of molecules are added in such a way (X- or Z-type) that the molecules in successive layers are all aligned in the same direction. Although this clearly provides the necessary acentric film structure, it is by no means always possible to achieve X- or Z-type deposition. A more generally useful approach has been to use alternating layers deposited in the Y-type mode, with the active layer interspersed with well-ordered layers of non-active amphiphiles. Although small amphiphiles can give highly ordered layers, the resulting films tend to be very fragile and prone to re-organisation. However the use of polymeric amphiphiles for both of the alternating layers has yielded structures composed of up to 300 bilayers which are stable at room temperature for at least several months[7]. The ability to build up large numbers of layers while preserving order is particularly important for obtaining high-efficiency SHG, since the resultant amplitude of the second harmonic is directly proportional to the number of layers, so that the intensity of the second harmonic depends on the square of the number of layers.

Use of poled polymers

Polymers rarely form good single crystals, but it is relatively easy to achieve a non-centrosymmetric alignment of NLO-active side groups on polymer chains even in non-crystalline material. Since the active side groups are dipolar, it is possible to heat the polymer above its glass transition temperature and apply a strong electric field to align all the polar side groups. The material is then cooled to below the glass transition temperature with the field still applied, to 'freeze-in' the aligned polar orientation. Although thermal motion occurs at the high temperature so that perfect alignment is not possible and the presence of the NLO-inactive polymer backbone effectively dilutes the active material limiting the overall magnitude of NLO effects, this is nevertheless a useful method. Such materials are beyond the scope of this text as they are not strictly crystalline so the reader is referred to a useful perspective overview[8]. NLO polymers are claimed to have many advantages, including fast response times, large NLO coefficients, low dielectric constants and switching energies, ease of processing, and good environmental, mechanical and structural stability.

In addition to the presence of a conjugated π-electron system, electron-donor and electron-acceptor moieties and, for second-order effects, absence of centres of symmetry in both the free molecule and its crystalline state, several other criteria must be satisfied for NLO applications:

(1) the material must be optically transparent at the laser wavelength and at the second-harmonic wavelength, as well as of good optical quality
(2) it must be possible to grow large high-quality single crystals or to obtain films having good crystalline order or a long single-crystalline core in a hollow optical fibre
(3) ideally the molecular orientation in the crystals should permit good phase-matching as described above
(4) the crystalline material should be chemically and physically stable.

All of these criteria are easier to achieve if the material is first purified to a high level by the methods described in chapter 1, as pure material will contain less of the defects discussed in chapter 4 and will in turn be less prone to the solid-state reactions discussed in chapter 7. The importance of crystal quality is illustrated by the fact that SHG efficiency is three times greater in the best crystal quality regions of a sample of 2(α-methylbenzylamino)-5-nitropyridine (MBA-NP) (figure 10.4) than in the worst regions[9]. The problems faced in growth of suitable crystals are well-illustrated by the case of DAN (figure 10.1) whose NLO properties were referred to above. Although this forms the necessary non-centrosymmetric

10.4. MBA-NP.

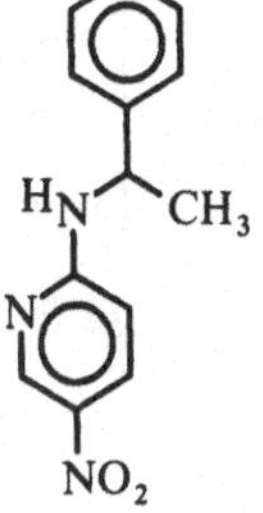

crystals, it is only slightly soluble in many solvents and decomposes slowly in the melt. Although slow evaporation of methanol solutions gives suitable seed crystals for further growth, the solubility ratio (defined as the ratio of the temperature-dependence of solubility, dS/dT, to the average solubility over the temperature range, S_a) is substantially lower than the optimum value of 0.01–0.03 K^{-1}, so that crystal growth is very slow, typically several months for a crystal with dimensions of a few millimetres. Fortunately, if highly purified material is used with melt-growth techniques in which the material is only in the molten state for the minimum time (achieved by using a thin molten interface between the growing crystal and the solid source material), good crystals of centimetre dimensions can be grown in reasonable time.

10.2   Third-order NLO materials

Most of the above discussion has concerned materials for second-order non-linear uses. Third-order non-linear organic materials have been much less extensively studied and are much less well understood. It is not necessary to achieve a non-centrosymmetric structure in this case, nor are electron-donor and electron-acceptor groups necessary. The largest class of materials studied in this context is the conjugated polymers such as

10.5.

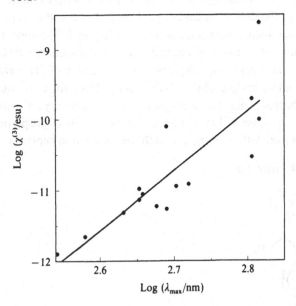

polyacetylene, polydiacetylene (c.f. chapter 7), poly-*p*-phenylenevinylene, etc. Their $\chi^{(3)}$ values range from about 10^{-12} to 10^{-8} cm^4 esu^{-2}, which is more than 10^7 times smaller than for the best inorganic materials. However, their responses are much more rapid and are virtually independent of laser wavelength, whereas use of the inorganic materials (such as GaAlAs) requires the use of wavelengths in the tail of the semiconductor band edge. There is currently some debate about whether significant improvements in the $\chi^{(3)}$ values for organic materials are in principle achievable to the necessary extent for them to be practically useful. A correlation between $\chi^{(3)}$ and the position of the lowest absorption maximum (figure 10.5) suggests that the ultimate limit may still be too small to be useful, whereas others have argued that wider exploration of the optimum operating conditions could yield higher values than those so far achieved and used in the correlation upon which this argument is based. Although further work is clearly necessary in this area, the prospects seem significantly less favourable than those for second-order applications.

Finally, some assessment of current progress towards device applications is relevant as this has been one of the main driving forces behind research in this area. The high performance of single crystals of optimised organic NLO materials has clear specialist laboratory applications such as frequency doubling and optical mixing, although the complications of growth of large organic crystals and the obvious possibility of long-term photochemical damage limit such applications. The use of films of organic materials in microscale opto-electronics seems more promising, and there have been reports of prototype electro-optic modulators using poled polymers[10]. Despite the limited actual uses of organic materials in devices at present, their potential justifies further work. Much of the initial work in this area was carried out by groups in which academic laboratories were linked with industrial companies, and the subject did not gain widespread exposure in the open literature until recently[e.g. 1,3,6,11–13]. The wider realisation by the scientific community of the opportunities will facilitate further developments, and the true potential of this adolescent field cannot yet be determined on the basis of current achievements in device applications.

References

1 S. Allen in *Molecular Electronics*, Ch. 4, ed. G.J. Ashwell, Wiley: Chichester, 1992.
2 C. Cassidy, J.-M. Halbout, W. Donaldson and C.L. Tang, *Opt. Commun.*, 1979, **29**, 243. J. Zyss, D.S. Chemla and J.F. Nicoud, *J. Chem. Phys.*, 1981, **74**, 4800. S. Wax, M. Chodrow and H.E. Puthoff, *Appl. Phys. Lett.*, 1970, **16**, 157.
3 D. Pugh and J.N. Sherwood, *Chem. in Britain*, 1988, **24**, 544.

4   S.T. Howard, M.B. Hursthouse, C.W. Lehmann, P.R. Mallinson and C.S. Frampton, *J. Chem. Phys.*, 1992, **97**, 5616.
5   Reference 1, p. 212.
6   J.F. Nicoud in *Organic Materials for Non-linear Optics*, ed. R.A. Hann and D. Bloor, Royal Society of Chemistry Special Publication No. 69, 1989, p. 157–162.
7   P. Hodge, *J. Mater. Chem.*, 1994, in press.
8   D.R. Ulrich in *Organic Materials for Non-linear Optics*, ed. R.A. Hann and D. Bloor, Royal Society of Chemistry Special Publication No. 69, 1989, p. 241–263.
9   J.N. Sherwood in *Organic Materials for Non-linear Optics*, ed. R.A. Hann and D. Bloor, Royal Society of Chemistry Special Publication No. 69, 1989, p. 71–81.
10 W.H.G. Horsthuis, F.C.J.M. van Veggel, J.-L.P. Heideman and C.P.J.M. van der Vorst, *Proc. EOA III*, 1992, p. 9/1–7, Plastics and Rubber Institute (ISBN 1874667101).
11 D.S. Chemla and J. Zyss (ed.), *Non-linear Optical Properties of Organic Molecules and Crystals*, Orlando: Academic Press, 1987 (2 volumes).
12 D.J. Williams (ed.), *Non-linear Optical Properties of Organic and Polymeric Materials*, Washington, DC: Am. Chem. Soc. Symp. Series 233, 1983.
13 R.A. Hann and D. Bloor (ed.), *Organic Materials for Non-linear Optics II*, Cambridge: Royal Society of Chemistry, 1991.

Index

Printed in the United States
By Bookmasters